INTRODUCTORY REMOTE SENSING: DIGITAL IMAGE PROCESSING AND APPLICATIONS

Paul J. Gibson and Clare H. Power

With Contributing Editors Sally E. Goldin
and Kurt T. Rudahl

ROUTLEDGE
ROUTLEDGE
Taylor & Francis Group

First published 2000
by Routledge
11 New Fetter Lane, London EC4P 4EE

Simultaneously published in the USA and Canada
by Routledge
29 West 35th Street, New York, NY 10001

Routledge is an imprint of Taylor & Francis

Typeset in Garamond by J&L Composition Ltd, Filey, North Yorkshire
Printed and bound in Great Britain by
St Edmundsbury Press, Bury St Edmunds, Suffolk

British Library Cataloguing in Publication Data
A catalogue record for this book is available from the British Library

Library of Congress Cataloging in Publication Data
A catalog record for this book has been requested

ISBN 0–415–18961–6 (hbk)
ISBN 0–415–18962–4 (pbk)

To my family, and in particular to Ian and Kay for sharing their wife and mother
with the rigours of authorship during 1997 and 1998

(CHP)

CONTENTS

PLATES

The following plates appear between pp. 124 and 125

2.1 (a) False colour MSS image; (b) corrected false colour image

3.1 TM 4 image shown (a) in its usual mode of varying shades of grey and (b) in density-sliced format
3.2 (a) False colour image of part of Peru; (b) a ratio image of same area
3.3 (a) False colour TM composite of part of Eritrea; (b) Intensity, Hue, Saturation image of the same region
3.4 False colour composite showing different types of vegetation and land use in western Ireland
3.5 Principal component versions of Plate 3.4
3.6 Principal component versions of Plate 3.2a
3.7 Synergistic image
3.8 Comparison of (a) TM image with (b) TM/SPOT fused data for an airport southeast of Los Angeles
3.9 Perspective view southeast of Los Angeles
3.10 Two false colour composite Landsat MSS images of Ireland
3.11 Density-sliced MSS 4 (infrared) image
3.12 (a) False colour composite; (b) training areas for supervised classification
3.13 (a) Supervised classification representation of Plate 3.12a; (b) feature–space plots for three two-band combinations
3.14 Unsupervised classification version of Plate 3.12a

4.1 ARTEMIS rainfall and cloud product
4.2 Normalised Difference Vegetation Index of Europe: an example of a global vegetation map

5.1 Sample of cloud type map for 1 July 1985 (0330)
5.2 Rainfall estimates for part of Africa for July 1985 for 10-day and 30-day periods
5.3 Normalised Difference Vegetation Index composite maps of Africa for January and July 1985 and 1986
5.4 (a) A false colour composite of Essex coast study area; (b) supervised maximum likelihood classification of land cover on the Essex coast
5.5 The surface model created by draping part of the Landsat TM image over the DEM
5.6 Effect of sea-level rise simulation of 2 m on the northern part of Landsat TM 3, 4, 5 image
5.7 An example of peatland ground cover, Wedholme Flow, Cumbria
5.8 The 25 November 1989 Landsat TM scene of the southern part of the study area in false colour

FIGURES

FIGURES APPEARING IN THE PRACTICAL MANUAL FOR DIGITAL IMAGE PROCESSING

TABLES

PREFACE
OUTLINE OF *INTRODUCTORY REMOTE SENSING*

Most of us have been introduced to remote sensing at some stage in our lives. At the shallowest level, this may only involve seeing a satellite image used as a 'pretty picture' adorning the frontispiece of a book or at a deeper level involving imagery of the planets of our Solar System transmitted by spacecraft such as Voyager. However, in seeking to obtain further information (how is the imagery transmitted? what dictates the observed colours? which is the optimum system? and so on), one can easily come away with the view that remote sensing is a complex subject composed of such a range of components that gaining an in-depth knowledge and understanding of it is impossible. *Introductory Remote Sensing* seeks to provide a greater understanding of the various aspects of remote sensing by considering this field of science in two volumes. This text is *Introductory Remote Sensing*, volume II (*Digital Image Processing and Applications*). It has a number of sections (Figure P.1). Section 1 explains how the digital data may be processed in order to maximise the information output. This section is divided into a brief introduction to remote sensing (Chapter 1), pre-processing the digital data (Chapter 2) and enhancing the imagery (Chapter 3). Section 2 provides a greater in-depth discussion on environmental monitoring techniques (Chapter 4) and case studies encompassing a range of disciplines are presented in Chapter 5. Digital image processing is a very important aspect of remote sensing and is conventionally considered within textbooks by showing the results of different procedures in plate format. However, it is essentially a practical hands-on process and a full understanding of the concepts can only be obtained by actually performing the processing

(section 3). The CD included in this volume includes a modified version of DRAGON image processing software for PC. This will allow the reader to:

- display images in single-band format and as three-band combinations in false-colour composites;
- zoom in to look at the images in more detail and to find out the digital values associated with specific positions on the images;
- produce DN histograms that summarise the distribution of the digital values in one band;
- produce scatterplots of two bands which provide a qualitative view of the correlation between bands;
- measure linear and areal features on the images with different units of measurement;
- perform contrast stretching, convolution filtering, ratioing and classification procedures.

In all, 77 separate datasets are supplied on the CD for nine areas, which include data from Europe, North America, South America, Africa and Asia. It is not possible to perform principal components analysis on the version of DRAGON that has been supplied but principal components images are also supplied for some of the datasets in order that they may also be investigated. In addition, Intensity, Hue and Saturation images are also included. The datasets comprise Landsat TM, SPOT and radar imagery. *Introductory Remote Sensing: Principles and Concepts* (the companion volume) is the basic remote sensing text and an outline of what is covered in this book is shown in Figure P.2. It is strongly recommended that *Principles and Concepts* be consulted because its contents are assumed knowledge for anyone using *Digital Image Processing*

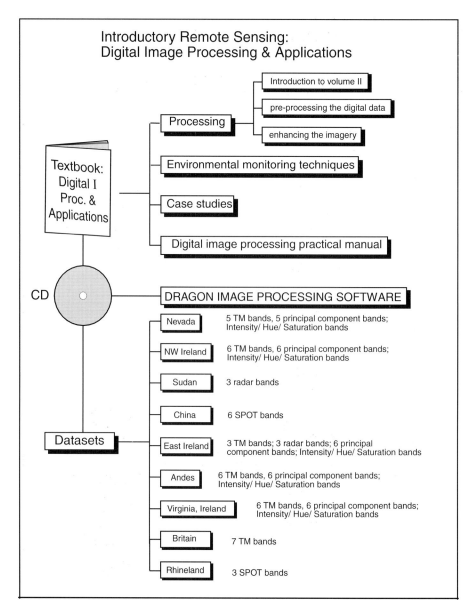

Figure P.1 Main components and layout of *Introductory Remote Sensing: Digital Image Processing and Applications*.

and Applications. Principles and Concepts seeks to address four main questions.

What is remote sensing?
What principles govern remote sensing?
How are remote sensing data obtained?

What use is remote sensing?

The companion volume also includes a dedicated WWW site that has a number of sections that can be accessed by Netscape version 3 (or higher) or Internet Explorer 4 (or higher). Additional image examples of

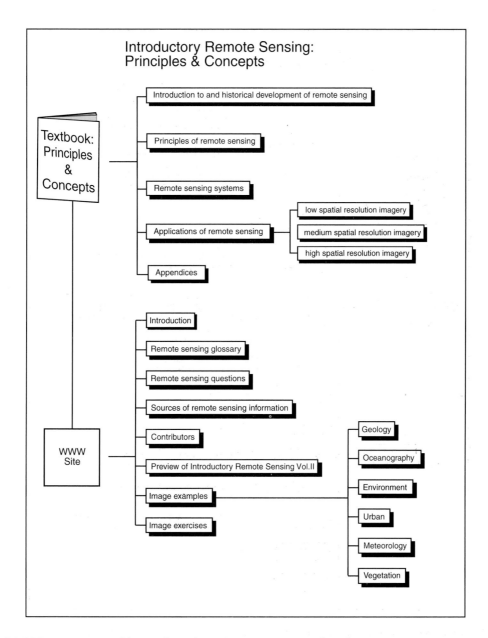

Figure P.2 Main components and layout of *Introductory Remote Sensing: Principles and Concepts.*

remote sensing for a number of important fields are presented: geology, meteorology, vegetation studies, urban studies, oceanography and environmental applications. An important component of remote sensing is the ability to interpret images. Conse- quently, a number of images are provided on the WWW site which may be used either in a classroom discussion context or as a means of determining your competence in image interpretation. Allied to this is a section containing questions (and answers) on remote sensing theory, covering all aspects of the companion volume. The questions require descriptive-type answers and simple mathematical calculations. Other sections within the WWW site include a glossary, sources of remote sensing information, a list of data contributors to the books and a preview of *Introductory Remote Sensing*.

The format for both books is similar and is structured in the following way:

- Each chapter concludes with a Further Reading section, though a much more comprehensive reading list is provided at the end of each book which allows readers to deepen their knowledge.
- A self-assessment test is included at the end of each chapter, in order that readers may determine whether they have fully grasped the concepts discussed within the chapter. Fully worked answers for all the questions are provided in appendices for both books.
- A number of boxes are included within the main text, e.g.:

A histogram for a non-enhanced image is shown in Figure 3.3. Calculate the output DNs in a contrast stretched image for the input values 60, 80 and 200.

The input range of 160 (200 − 40) is mapped to an output range of 0–255. Therefore each input unit is equivalent to 255/160 output units. The value 40 (lowest) in the input range is assigned a DN of zero (lowest) in the output, therefore:

60 in the input is equal to $20 \times (255/160) = 32$

80 in the input is equal to $40 \times (255/160) = 64$

200 in the input is equal to
$$160 \times (255/160) = 255.$$

These boxes relate to the theory that is currently being explained and generally take the form of examples, including worked answers, which allow readers

to be confident that they have fully understood a particular point.

- Shaded boxes are also included within the body of the text, e.g.:

Polynomial Transformation

A polynomial is essentially a mathematical equation which links the uncorrected image to the georeferenced database based on the input ground-control points. The polynomial dictates the translation, scaling and/or rotation that the uncorrected image has to undergo in order to be georectified. A polynomial equation may be of any order . . .

These shaded sections represent more complex components which may be skipped if such in-depth knowledge is not required or read in order to get a deeper insight. Both *Introductory Remote Sensing* volumes may be used, either by an individual who wishes to gain a full understanding of the various concepts or in a classroom format. *Principles and Concepts* provides an excellent introduction to the subject and is particularly suited for fifth grade to first-year university level. *Digital Image Processing and Applications*, which is more advanced and requires a level of computer experience, may be employed from sixth grade to third-year university level. The combination of the theory provided by the textbooks, exercises and applications on the dedicated WWW site and the practical component provided by the image processing capabilities in the CD accompanying *Digital Image Processing and Applications* make this entire package particularly suitable for teachers and lecturers who want to run a remote sensing course. With the increased application of remote sensing techniques in the fields of geography, geology, environmental monitoring, urban studies, oceanography and vegetation studies, the acquisition of remote sensing skills is extremely desirable and is expected to become increasingly important in the future.

ACKNOWLEDGEMENTS

The authors would like to thank Mr J. Keenan, Hilary Foxwell and Pat Brown for producing the figures in this book. A book such as this relies heavily on the goodwill of many individuals and organisations to provide relevant remote sensing imagery and data and the permission to use them. These organisations are listed in Appendix E but especial thanks go to ERA-Maptec, SPOT IMAGE, National Remote Sensing Centre Ltd, EUMETSAT, Plymouth Marine Laboratories, Cambridge University Committee for Aerial Photography, UK Defence and Environmental Research Agency, Ordnance Survey, European Space Agency, US National Oceanic and Atmospheric Administration and NASA. The permission to use DRAGON software by Sally Goldin and Kurt Rudahl of Goldin–Rudahl Systems Inc. was greatly appreciated.

The case studies in particular have been drawn from projects carried out with colleagues and the work has been achieved by team effort. Clare Power would therefore like to acknowledge a number of colleagues, past and present, for their advice and contributions to these projects: Eric Barrett, Andrew Harrison, Giles D'Souza, Daniel Lloyd, Chris Kidd, Richard Lucas, Tim Richards and Mike Beaumont from the University of Bristol Remote Sensing Unit; Nick Veck, Neil Hubbard, Debbie Needham, David Fox, Mark Richardson and Richard Knowles from the National Remote Sensing Centre Ltd, Farnborough; Ian Downey, Anita Perryman and Stephane Flasse from the Natural Resources Institute, University of Greenwich; and Alistair Baxter, Head of Department, School of Earth and Environmental Sciences, University of Greenwich; for the many facilities he provided. Many clients and collaborators have provided the requirement or reason for many of these projects and the funding and/or data for them. The Sultanate of Oman Public Authority for Water Resources and the Council for the Environment and Water Resources for the Sultanate of Oman, the UK Natural Environment Research Council, the United Nations Food and Agricultural Organisation, the US National Oceanic and Atmospheric Administration, the EU, the UK Wild Fowl and Wetlands Trust, Air Photo Services Cambridge, Lancaster University Archaeological Unit, UK Defence Research Agency (DERA) and the Ordnance Survey.

A grant from the Publications Committee of the National University of Ireland, Maynooth, was gratefully appreciated (PJG).

1

INTRODUCTION TO DIGITAL IMAGE PROCESSING AND APPLICATIONS

Chapter Outline

1.1 Introduction

1.2 Passive remote sensing systems

1.3 Reflectance characteristics of landscape features

1.4 Active remote sensing systems

1.5 Image analysis

1.6 Digital images

1.7 Chapter summary

Self-Assessment Test

Further Reading

1.1 INTRODUCTION

This book is the second volume of *Introductory Remote Sensing* and concentrates on the various digital image processing techniques that can be applied to remotely sensed datasets and the applications of remotely sensed images. *Introductory Remote Sensing: Principles and Concepts* dealt with the principles and concepts behind remote sensing and a good knowledge of these principles is required in order fully to understand and exploit the contents of this book. This chapter forms a very short précis of *Principles and Concepts*. The reader is strongly encouraged to read *Introductory*

Remote Sensing: Principles and Concepts for a more detailed explanation of the terms introduced here, or one of the general reference texts included at the end of the chapter.

Remote sensing can be defined as the acquisition and recording of information about an object without being in direct contact with that object. Remotely sensed images obtained in the early part of the twentieth century consisted of aerial photographs that were generally within the visible range of the electromagnetic spectrum. The terms 'photograph' and 'image' are not synonymous. A photograph is a record of a scene captured on film. Current photographic film is sensitive from the high ultraviolet to the photographic infrared part of the electromagnetic spectrum. An image is a record of a scene that may have been captured on film or obtained by a scanner system. Although photographic film continues to form an important component of remote sensing, it has to a large extent been superseded, especially for satellites, by scanner systems which record data in a digital format. The use of a scanner system allows the acquisition of data at wavelengths longer than visible light such as thermal infrared or in the microwave range. Since the 1960s, remote sensing platforms have operated from orbit and have performed many functions. Polar-orbiting satellites such as Landsat, SPOT, Indian Remote Sensing (IRS) satellite and JERS-1 (Japanese Earth Resources Satellite) have spatial resolutions of less than 100 m and have applications in environmental monitoring, whereas the polar-orbiting NOAA satellites have a much coarser resolution but image a very wide swath and are primarily designed for meteorological purposes. Other meteorological satellites such as Meteosat or GOES operate in geostationary orbits (36,000 km) and image very large segments of the globe. Remote sensing systems may be either passive or active.

1.2 PASSIVE REMOTE SENSING SYSTEMS

Passive systems measure electromagnetic radiation that is reflected or emitted from a surface. The primary source of the electromagnetic radiation for a passive system is the Sun. The Sun does not produce a constant amount of energy across all wavelengths; most is produced in the visible and infrared ranges with very little in the shorter X-ray and gamma ray range or the longer microwave range. Although electromagnetic radiation of all wavelengths reaches the outer atmosphere, only some wavelengths reach the surface of the Earth. The atmosphere is composed of gases that selectively absorb electromagnetic radiation of particular wavelengths. Wavelengths shorter than the ultraviolet are absorbed. A number of atmospheric windows exist at particular wavebands through which radiation may pass, sensors onboard

remote sensing systems are thus designed to obtain their data at these specific wavebands. (Some sensors are specifically designed to obtain information on the atmosphere, such as water vapour content, and these operate at wavelengths at which water vapour absorbs the radiation). A broad absorption band exists between 14 and 17 μm and also at 2.7 μm and 4.5 μm because of carbon dioxide. Water vapour absorbs at 1.4 μm, 2.7 μm and 6.3 μm. The most important atmospheric windows are in the visible/near infrared and microwave ranges and in parts of the thermal infrared (3−5 μm and 8−14 μm). As well as selectively absorbing particular wavelengths of electromagnetic radiation, aerosols in the atmosphere scatter electromagnetic radiation. Selective Rayleigh or Mie scattering is wavelength dependent where Rayleigh scattering is inversely proportional to the fourth power of wavelength and Mie scattering is inversely proportional to wavelength. Shorter wavelengths are therefore affected to a greater degree than longer ones. Thus, when the reflectance from a scene is recorded, a component of the measured signal is due to scattering in the atmosphere and this atmospheric contribution varies according to the wavelength at which the measurements were taken.

Typical passive remote sensing platforms are Landsat and SPOT. Landsat MSS (multispectral scanner) obtains data at four wavebands (green to near infrared) with an 80 m spatial resolution whereas Landsat TM (Thematic Mapper) obtains data in six reflected bands (30 m spatial resolution) and one thermal band (120 m spatial resolution). SPOT, a French satellite series, the first of which was launched in 1985, can obtain data in two modes. In multispectral mode, it obtains data in three wavebands with a spatial resolution of 20 m and in panchromatic mode with a 10 m spatial resolution. (SPOT 4, launched in 1998, obtains four bands of data in multispectral mode). Although Landsat and SPOT are passive systems, they obtain their data differently. Landsat uses a transverse (across-track) scanning system that consists of a mirror which rotates through a small angle and directs the reflectance from the terrain onto banks of detectors where the strength of the signal is measured in specific wavebands. SPOT is a pushbroom system which employs a very large number of charge-coupled devices (3,000 in panchromatic mode, 6,000 in multispectral mode) and each detector measures the reflectance for a small area termed the 'ground-resolution cell'. SPOT has the added advantages of being able to obtain data from regions not directly beneath the satellite (off-nadir) and it is possible to view a pair of SPOT images stereoscopically.

1.3 REFLECTANCE CHARACTERISTICS OF LANDSCAPE FEATURES

Most satellite-based passive remote sensing systems obtain data in the visible and infrared and a knowledge of the reflectance characteristics of vegetation, soils, water and rocks at these wavelengths is necessary in order to interpret an image correctly. Although we perceive vegetation as green (i.e. in the visible range it has a higher reflectance in green than in either the blue or red range), the reflectance within the near infrared, is substantially greater than for the green range. Plant species can often be differentiated in the infrared range because of a marked difference in their reflectances compared with their reflectances in the visible. This property of vegetation, a low reflectance in the red and a high reflectance in the infrared, is employed when vegetation indices are being produced by the use of ratio images, which are discussed in Chapters 3 and 4 of this volume. Soils generally have very low reflectance in the visible and infrared. Rocks are formed of minerals and the spectral characteristics of these minerals influence the resultant signature of the rocks. A mineral which contains iron may have an absorption band in the $0.9-1.0$ μm range if it is in the ferrous state and at 0.7 μm if it is in the ferric state. Carbonates may produce characteristic absorptions around 2.2 μm and hydroxyl ions, which are often associated with clays, yield absorptions at 1.4 μm, 2.2 μm and 2.3 μm. Although it is often possible to determine the existence of different rocks in an area (basalt is generally dark and granite is much paler), it is generally not possible to assign a specific signature unequivocally to a particular rock type. The reflectance characteristics of water vary greatly depending upon whether it is in the liquid or solid phase. Snow and, to a slightly lesser degree, ice have a very high reflectance whereas the reflectance for water is low in the visible range and very low in the near infrared. Thus it will appear dark on an infrared image. However, at low illumination angles, water may act as a specular reflector and reflect a large proportion of the incident energy, producing a bright signature.

1.4 ACTIVE REMOTE SENSING SYSTEMS

The second form of remote sensing is active remote sensing, in which electromagnetic radiation of a specific wavelength is generated by the system. This is directed at the surface and the energy that is scattered back from the surface is recorded. The most common form of active remote sensing is radar, which operates within the microwave range of the electromagnetic spectrum. Unlike passive systems, active radar systems do not image: the area directly beneath the aircraft or satellite but illuminate the terrain to the side. The radar system requires two pieces of information to produce a radar image: time and the strength of the returned signal. The time delay between transmitting the pulse of electromagnetic radiation and recording the returned signal after it has been scattered back from an object allows the distance of the object from the remote sensing system to be calculated. The strength of the returned signal dictates the tone that will be assigned to the object. If a large proportion of the energy is returned, the object is assigned a bright signature whereas if very little energy is scattered back to the platform then a very dark signature is recorded. The parameters that control a radar image include polarisation, slope aspect, dielectric constant, wavelength and surface roughness.

Although the microwave range varies from 0.1 cm to 100 cm, most radar systems operate at a specific wavelength such as 23 cm. Long-wavelength radars can penetrate cloud, and are thus particularly useful for investigation in tropical regions where cloud cover can be persistent. Short-wavelength radars have meteorological applications. Both short- and long-wavelength radars, because they provide their own source of illumination, may operate at night. The roughness of a surface can be determined by the modified Rayleigh criterion. Radar-rough surfaces yield high radar returns because a large proportion of the incident energy is scattered back towards the sensor whereas a radar-smooth surface reflects the incident energy away from the platform and a dark signature is recorded. Surface roughness is measured at the centimetre scale. Surfaces that are sloped may also yield a bright signature.

A number of radar satellites are currently in operation. RADARSAT, a Canadian system, operates at a 5.6 cm wavelength. It has a number of operating modes, allowing the acquisition of different swath widths with different spatial resolutions. Two radar satellites (ERS-1 and ERS-2) launched by the European Space Agency also operate at 5.6 cm. The electromagnetic pulse generated by radar systems is polarised and usually designed to vibrate in the vertical or horizontal plane. When it strikes a target, a proportion of the energy is depolarised and the radar system may measure the returned pulse with the same orientation as the emitted pulse or the pulse that is polarised at right angles to the emitted one. RADARSAT transmits and receives in the horizontal plane whereas ERS operates in the vertical plane. Those radar systems that are placed on spaceborne platforms (and the majority of airborne platforms as well) are Synthetic Aperture Radars (SARs). Such systems give excellent azimuth resolutions by measuring the frequency change for an object as it moves through the radar beam (Doppler shift). A growing trend for satellites that have been placed in orbit in recent years has been to combine both passive sensors and an active radar system on the same platform. The Japanese satellite JERS-1 carries sensors similar to SPOT but also has on board a 23 cm wavelength SAR.

1.5 IMAGE ANALYSIS

Image interpretation involves distinguishing different surface classes based mainly on tonal and textural differences. However, the context and spatial relationships of various elements of the image are also important in image interpretation.

False colour images are used extensively in remote sensing. One can consider a true colour image as being formed by combining:

- an image obtained in the blue range of the spectrum projected in blue light,
- an image obtained in the green range projected in green light,
- an image obtained in the red range projected in red light.

However, if, for example, two surfaces have similar reflectances in the blue range but different in the infrared, an image produced from green, red and infrared data may allow a greater degree of differentiation. A false colour image can be produced by combining:

- an image obtained in the green range of the spectrum projected in blue light,
- an image obtained in the red range projected in green light,
- an image obtained in the infrared range projected in red light.

Vegetation, because of its very high reflectance in the infrared, often produces a characteristic red signature in a false colour image.

The general term 'lineament' is applied to linear topographic tonal or textural features that are observed on an image. Many of these lineaments are cultural features such as roads or runways, though natural features in the landscape may also produce them. Natural lineaments are important in geological investigation because they may represent unknown faults or fractures. Major faults may act as conduits for the movement of mineralised fluids and may thus be associated with economic mineral deposits.

1.6 DIGITAL IMAGES

A digital image, obtained by a scanner system, is a regular grid array in which a digital number (DN) is

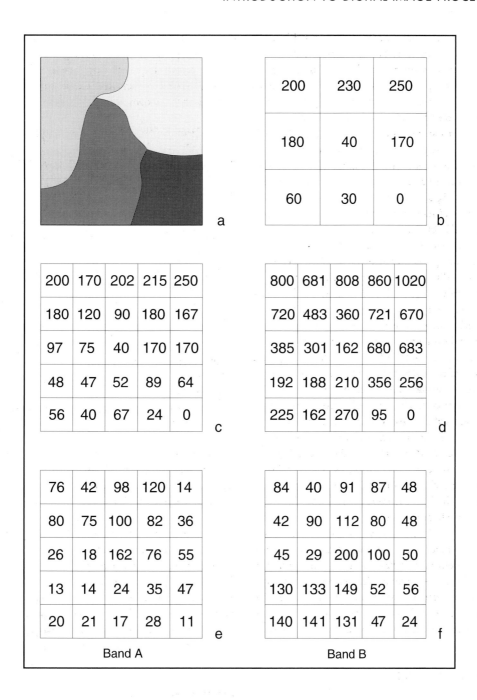

Figure 1.1 (a) Simple image and its corresponding digital numbers for (b) low spatial resolution 8-bit system; (c) high spatial resolution 8-bit system; (d) high spatial resolution 10-bit system; (e) and (f) high spatial resolution, 8-bit, two-band system.

assigned by a remote sensing system's detector, which is related to the parameter that is being measured. Figure 1.1a shows a hypothetical part of the Earth that is characterised by different reflectances, in which the lower reflectance areas are illustrated by darker tones. A digital image representation of this scene is given in Figure 1.1b, in which each square, termed a pixel, is assigned a DN. A comparison of Figure 1.1a and 1.1b shows that, for this digital image, lower reflectances are associated with lower digital numbers. This convention is followed by remote sensing systems such as Landsat and SPOT. A second digital image representation of the scene obtained by a different remote sensing platform is shown in Figure 1.1c. When it is compared with Figure 1.1b; it can be seen that it has many more digital numbers (each associated with a pixel) than the latter. Each pixel in Figure 1.1c represents a smaller area than a pixel on Figure 1.1b, thus the system which obtained Figure 1.1c has a better spatial resolution than the remote sensing system that obtained Figure 1.1b. A 60×60 km area on a single-band digital image produced by SPOT with a spatial resolution of 20 m typically has 9 million pixels whereas a TM scene of the same area, with a 30 m spatial resolution, has 4 million pixels. Although, at first glance, Figures 1.1 c and 1.1d appear similar as they both have the same spatial resolution, the range of digital numbers is much larger for the latter. The remote sensing systems that obtained the data in Figures 1.1b and 1.1c both have an 8-bit radiometric resolution and the grey levels in the original scene are expressed in 256 levels ($2^8 = 256$). However, the better 10-bit radiometric resolution for Figure 1.1d allows more grey levels to be measured. (Note how the digital numbers of 170 in Figure 1.1c are different in 1.1d, showing that the improved radiometric resolution of the latter has allowed a better representation of the original scene). It is important to understand that DNs are arbitrary units which are used to indicate variations in radiance. Absolute radiance values may be obtained from the DNs by means of calibration graphs.

Most remote sensing systems are multispectral and obtain data simultaneously in specific wavebands.

Landsat TM, for example, obtains data in seven bands; thus a TM scene would have seven digital images associated with it. The spectral resolution of a remote sensing system is a measure of how narrow the wavebands are and how many of them are recorded. Figures 1.1e and 1.1f illustrate the digital images obtained by a remote sensing system which has a similar spatial and radiometric resolution to that which acquired 1.1c, but in two bands. Note how the DN for the bottom left-hand corner pixel (row 5, column 1) is markedly different for band A and band B. Such a situation may arise if the area is vegetated and band A is measured in the red waveband and band B in the photographic infrared range. In addition, the pixels in row 1, column 5 and row 2 column 5 cannnot be distinguished on band B but can be on band A. Digital data from multiband systems are necessary for particular image processing procedures such as ratioing.

Digital data have a number of advantages over photographic film. Measurements (such as variations in reflectance) can easily be recorded digitally but – more importantly – the data can be transmitted at the speed of light (3×10^8 m/s) from a satellite platform to a receiving station. The satellite and receiving station do not need to be in a direct line of sight because the data can be transmitted through geostationary relay satellites. Digital data can be copied repeatedly without any loss of data because they are retained in a numerical format. Thus a copy of the data is identical to the original. However, if a copy of a photograph is itself further copied, the copies are degraded when compared to the original. This ability to transmit data quickly without any loss is particularly important, because today researchers from different parts of the world often collaborate on remote sensing projects. Digital data can be electronically mailed in order that all those involved in the research have access to the same dataset. Today datasets are often sited on open-access central servers and interested parties may download them. An advantage that photographic media have historically enjoyed has been their ability to store large amounts of information compared with digital media. Gillespie (1980) reported that a 35 mm black and white slide can contain more than 2×10^8 bits of information. Even 10 years ago, a number of reels of computer tapes would have been required to store so much data digitally, though CDs currently have capacities of around 600 megabytes. An important advantage of a digital data format is that it lends itself very easily to mathematical manipulation by digital image processing techniques. Such procedures enhance the images and allow a greater degree of information extraction. Photo-optical enhancement techniques are available (Skaley 1980) but a greater degree of control is possible if digital manipulation is used. In order to maximise the information output from the data obtained by remote sensing systems, it is necessary to process the data. The types of processing that may be performed on the data are considered in Chapters 2 and 3 and applications are discussed in Chapter 5.

1.7 CHAPTER SUMMARY

- Remote sensing systems mainly operate in the visible, infrared or microwave parts of the electromagnetic spectrum.
- Polar-orbiting satellites generally have a relatively high spatial resolution (less than 100 m) and image a relatively narrow swath (less than 200 km) or have a coarse spatial resolution but image a wide swath of terrain.
- Remote sensing systems are passive if they simply record the radiation that is reflected or emitted from a surface, and active if they carry their own electromagnetic source. SPOT and Landsat are passive, whereas radar systems such as RADARSAT and ERS are active.
- Landscape features have different reflectance properties in different ranges of the electromagnetic spectrum and may therefore appear different on remotely sensed images. Thus different vegetation or rock types can be distinguished from each other.
- Most spaceborne systems obtain their data in a digital format from which images can be produced.
- The spatial resolution is a measure of how much detail can be observed. The radiometric resolution is a measure of how many grey levels are measured. Many systems are multispectral: a number of bands of data for the same area are recorded but in different parts of the electromagnetic spectrum.

SELF-ASSESSMENT TEST

1 List three polar-orbiting satellites which have spatial resolutions less than 100 m and one polar-orbiting satellite with a much coarser resolution.

2 Why is the 8–14 µm range important in remote sensing?

3 Which gaseous components of the atmosphere absorb radiation at 2.7 µm?

4 How does Landsat differ from SPOT?

5 What are the important parameters in radar imaging?

6 How many grey levels does a 10-bit system measure and what is the highest DN that can be recorded?

7 Why are false colour images used in remote sensing?

FURTHER READING

Campbell, J. B. (1996) *Introduction to Remote Sensing*, London: Taylor and Francis (Chapters 2 and 3).

Drury, S. A. (1993) *Image Interpretation in Geology*, London: Chapman and Hall (Chapters 1 and 3).

Drury, S. A. (1998) Images of the Earth: *A guide to remote sensing*, Oxford: Oxford Science Publications (Chapters 2 and 3).

Gibson, P. J. (2000) Introductory *Remote Sensing*, volume I: *Principles and concepts*, London: Routledge (Chapters 2 and 3).

Gillespie, A. R. (1980) 'Digital techniques of image enhancement', in B. S. Siegal and A. R Gillespie (eds) *Remote Sensing in Geology*, New York: John Wiley and Sons Inc.

Lillesand, T. M. and Kiefer, R. W. (1994) *Remote Sensing and Image Interpretation*, New York: John Wiley and Sons (Chapters 5 and 6).

Sabins, F. F. (1997) *Remote Sensing: principles and interpretation*, New York: W. H. Freeman and Company (Chapters 4 and 6).

Skaley, J. E. (1980) 'Photo-optical techniques of image enhancement', in B. S. Siegal and A. R. Gillespie (eds) *Remote Sensing in Geology*, New York: John Wiley and Sons Inc.

2

PRE-PROCESSING THE DIGITAL DATA

2.1 INTRODUCTION

Before the data obtained by satellite systems or digital airborne systems can be analysed, it is often necessary to pre-process them in order to correct defects. Errors may arise from a number of sources:

1 Defective detectors in the scanning system may result in the acquisition of spurious data.
2 Data may be lost during transmission from the satellite to a ground-based receiving station.
3 Data may be corrupted during recording onto computer-compatible tapes at the receiving station.

4   There may be scale distortions of the images.
5   The measured digital number (DN) may contain an atmospheric component.

Whatever the source of the errors, data may be rendered totally or partially unusable without proper pre-processing. This pre-processing is often referred to as 'cosmetic processing', as the resultant image is more visually appealing. Figure 2.1a shows a TM 5 image which appears quite sharp and devoid of errors. A TM 6 image of the same area (Figure 2.1b) obtained simultaneously, but using different detectors, is clearly defective. The horizontal banding is not related to any characteristics of the scene but to problems in the acquisition of the data. This example illustrates an important point, that errors in data collection may be associated with only one band of data. Thus one should not assume that, because the TM 5 image in Figure 2.1a appears to be free of errors, all other bands also contain good data. It is necessary to examine all bands individually for errors. A useful concept in visualising some of the corrections that are applied to digital data is that of the digital image histogram.

2.2 DIGITAL IMAGE HISTOGRAM

Many digital image processing techniques involve substituting the digital number (DN) associated with a particular pixel with another DN in order to enhance a particular feature. However, in order to optimise this substitution or to ascertain whether it is even necessary, it is important that the distribution of the DNs in the scene is known. Most digital image processing systems have the capability to produce a table which shows the frequency of occurrence of every DN. A table of digital numbers is not very instructive for visualising the DN distribution in a scene. However, a DN table can be easily converted into a frequency histogram (also referred to as a DN histogram) which is a graphical representation of the dataset. The list of numbers shown in Figure 2.2 does not convey an impression of the DN distribution whereas the histogram formed from the data clearly shows the distribution is not symmetrical. The x-axis of such a histogram shows the DNs which are present while the height of the bars ('DN bin') indicates the number of pixels which have a particular DN. In the example shown, all DN bins less than 64 and greater than 85 are empty. For an n-band system, n histograms can be produced; a Landsat TM image will

 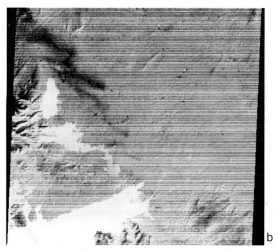

Figure 2.1 (a) TM 5 image produced by correctly operating detectors. (b) TM 6 image of the same region obtained at the same time as the TM 5 image. Defective detectors have produced horizontal banding.

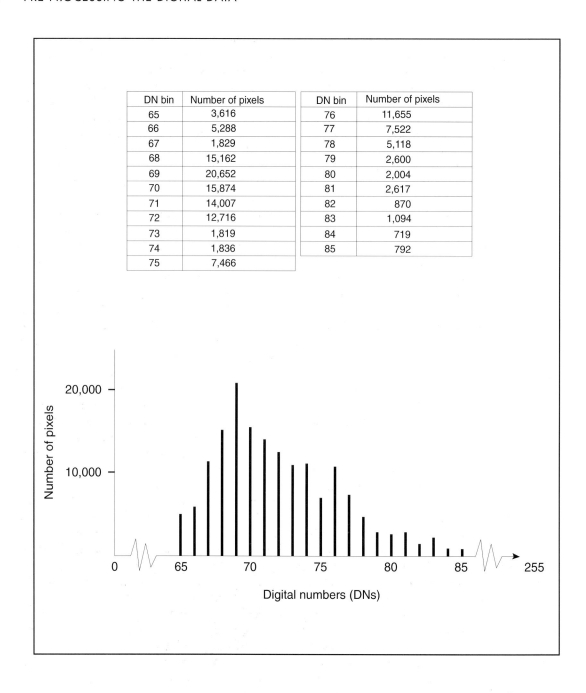

DN bin	Number of pixels
65	3,616
66	5,288
67	1,829
68	15,162
69	20,652
70	15,874
71	14,007
72	12,716
73	1,819
74	1,836
75	7,466

DN bin	Number of pixels
76	11,655
77	7,522
78	5,118
79	2,600
80	2,004
81	2,617
82	870
83	1,094
84	719
85	792

Figure 2.2 DN histogram produced from table of data. The graphical representation of the histogram allows a better understanding of the distribution of the data.

thus have seven histograms. In general, the histogram is not computed by means of all the DNs in a scene; a full TM image would require approximately 266 million digital numbers to be read. Instead, a representative sample is used in the computation. Some image processing systems allow the operator to introduce a 'skip factor' in the derivation of the histogram. Thus every nth pixel and/or nth line is used. A skip factor of 2 for both pixels and lines uses the data from every second pixel on every second line thus: (1,1); (1,3); (1,5) . . . (3,1); (3,3); (3,5) . . . etc. A skip factor of 5 uses 4 per cent of the data to construct the histogram. This approach allows regular sampling of the entire scene and would produce a better representative histogram than simply taking a small subscene in one area and assuming that the histogram produced from the subscene is representative of the entire scene.

The histograms for three SPOT bands for a scene in temperate terrain dominated by vegetation are shown in Figure 2.3. The shape of the histogram depends on the types of surfaces that are present, the extent of the surfaces and the wavelengths at which the reflectances are obtained. The band 1 (XS 1) and band 2 (XS 2) histograms appear similar though the peak for band 2 is lower (DN value of 31) than the peak value in band 1 (DN of 43) because vegetation has a higher reflectance in green (band 1) than in red (band 2). The histogram for band 3 is unlike the other bands. The digital numbers are much higher because of the very high reflectance of vegetation in the infrared. In addition, the histogram can clearly be considered to have two distinct sections. The broad plateau extending from 52 to 190 is due to the dominance of vegetation in the scene but the small number of pixels with low numbers clustered around a DN of 16 is due to the presence of water in the scene, which has a very low reflectance in the infrared.

Although every scene is in some sense unique, a number of generalised histogram forms can be recognised and a range of statistical measures used to describe the DN distribution (Figure 2.4). Many histograms have a tendency to cluster around a central DN value where the maximum numbers of pixels occur, with the numbers of pixels occupying each bin

decreasing away from this central DN in a symmetrical manner (Figure 2.4a). The simplest measure of this central DN is the mean, which is the arithmetic average, obtained by dividing the sum of all pixel values by the number of pixels. (Other statistical terms employed include the mode, which is the DN that occurs most frequently, or the median, which is the DN value midway in the frequency distribution.) The plot shown in Figure 2.4a is a normal frequency distribution and in such a situation the mean, mode and median are equal. The mean alone is not sufficient to describe how the DNs vary for a symmetrical distribution, as a comparison of Figures 2.4a and 2.4b shows. Both histograms have the same mean but the data are clustered more tightly about the mean in Figure 2.4b than in Figure 2.4a. A measure of the spread or the dispersion of the data about the mean is given by the standard deviation of the data. The term 'variance' is also used and it is equal to the square of the standard deviation. Not all DN histograms are symmetrical: they are often skewed, as shown in Figure 2.4c. In such instances the mean, mode and median of the dataset do not coincide. It is possible, using simple equations, to calculate the degree of skewness, which is a measure of how far the distribution departs from a normal distribution. Although the statistical mean and standard deviation may be calculated for the DN comprising any image, the numerical values obtained may not be particularly informative. The statistical mean for Figure 2.4d is similar to Figure 2.4a but it does not coincide with any digital number that occurs in the image. In this instance it may be more useful to calculate the means for each of the clusters of data.

The corrections discussed below apply to airborne and spaceborne systems and include:

- co-registration of data,
- line banding corrections,
- line dropout corrections,
- geometric corrections,
- atmospheric corrections,
- solar illumination corrections.

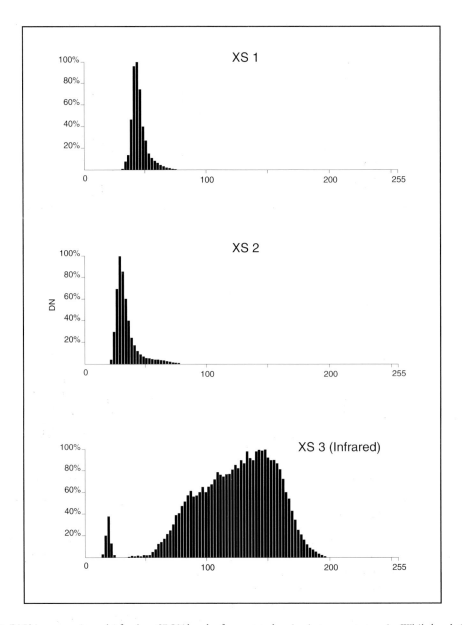

Figure 2.3 DN histograms (to scale) for three SPOT bands of a vegetated region in temperate terrain. While bands 1 and 2 are relatively similar (though band 1 has a higher mean than band 2), the DN distribution for band 3 is markedly different.

2.3 CO-REGISTRATION OF DATA

Multispectral datasets consist of pixels, each of which represents a specific area within a scene. For an n-band system, each pixel has n digital numbers. Before any processing can begin, it is important to ensure that the n bands are perfectly co-registered to each other, in order that the correct digital numbers are

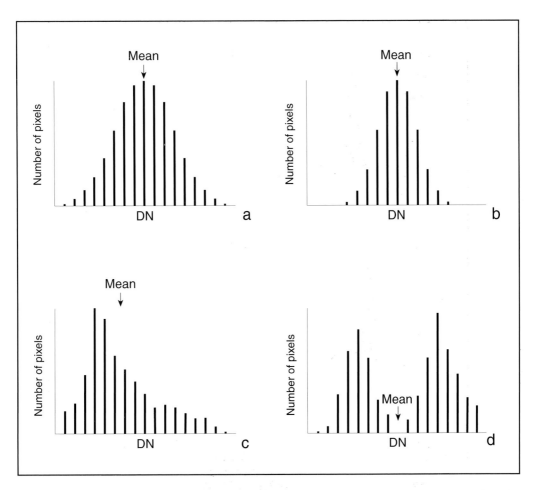

Figure 2.4 (a) Symmetrical DN histogram; (b) DH histogram similar to (a) but with a smaller variance; (c) skewed DN histogram; (d) bimodal distribution.

assigned to each pixel. Plate 2.1a shows a false colour MSS image produced by projecting band 1 in red, band 2 in green and band 4 in blue. There is a very prominent red 'ghost boundary' for the coast offset from the rest of the image. An examination of the dataset (which was obtained in this format from an international distributor) shows that the data for band 1 are offset eastwards by 178 pixels compared with the other bands. Thus for the false colour composite in Plate 2.1a, the digital numbers for the pixel at column 179, row 1, in bands 2 and 4 were superimposed with the digital number for the pixel at line number 1, row number 1 in band 1. Before further

processing, the band 1 data need to be shifted westwards until all the bands are properly co-registered (Plate 2.1b).

2.4 LINE-BANDING CORRECTIONS

Transverse scanning systems which use an oscillating mirror to obtain data usually employ more than one detector per band. Landsat MSS employs six detectors for each of the four sensed bands whereas Landsat TM has sixteen detectors for TM 1–5 and TM 7 and four detectors for TM 6. Consider the example shown in

Figure 2.5, in which a remote sensing system obtains three bands of data with four detectors per band. Detector 1 reads lines 1, 5, 9 ... detector 2 reads lines 2, 6, 10 ... detector 3 reads lines 3, 7, 11 ... and detector 4 reads lines 4, 8, 12 ... etc. Before launch, all the detectors for all bands are calibrated and their responses matched. Thus in theory all detectors for any particular band, if they analysed the same pixel, should each assign it the same digital number. However, satellite systems operate for many years and over time the detectors degrade and the response for all the detectors is not constant. (An everyday example of this is the situation where four watches are synchronised to show the exact same time. One year later, if the watches are again examined, they are likely to show slightly different times.) If one detector's

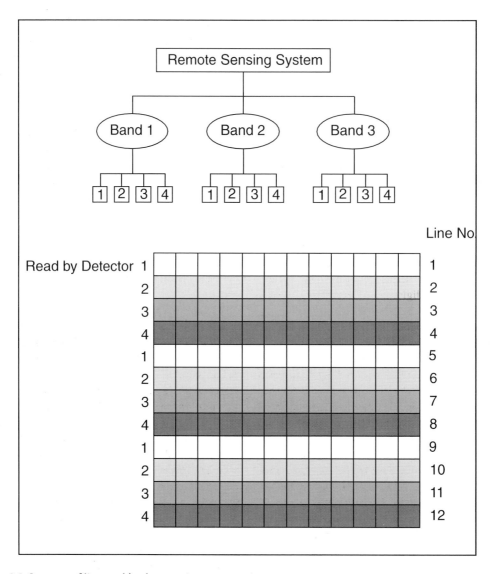

Figure 2.5 Sequence of lines read by detectors in remote sensing system.

response drifts to a much greater degree than the others, it will record digital numbers that are consistently higher or lower than those recorded by the other detectors. This manifests itself on an image as banding. The repeat cycle for the banding is a function of the number of detectors: Landsat MSS images, for example, may exhibit sixth-line banding if one detector is not synchronised with the others. It is possible to reduce the banding effect by a destriping process involving a number of steps.

1 A histogram for each detector in each band is produced. For the example in Figure 2.6, histogram 1 is formed from data in line 1, 5, 9 ... etc., histogram 2 from lines 2, 6, 10 ..., histogram 3 from lines 3, 7, 11 ... and histogram 4 from lines 4, 8, 12 ... etc.
2 Assuming each detector has sensed a representative sample of all the surface classes within the scene, each of the histograms will be similar (i.e. have the same mean and standard deviation) if the detectors are matched and calibrated. However, if one detector is no longer producing data readings consistent with the other detectors, its histogram will be different. Detector 4 in Figure 2.6 is producing higher digital numbers than the other detectors.
3 An average histogram is produced by using the digital number from all the detectors.
4 The DNs produced by all the detectors are altered so that their histograms are then made to match the average one. When this procedure is finished, the imbalance between the detectors is eliminated and the image is said to have been destriped. This procedure changes the DN for all lines, though the relative change for the properly functioning detectors is less for systems which have more detectors. A defective detector on the Landsat MSS forms one-sixth of the input to the average histogram whereas a defective detector for a reflected TM band contributes only one-sixteenth of the input to the average histogram. Figure 2.7a shows an MSS image with evidence of banding. After correction, the banding effect is reduced (Figure

2.7b). Line banding may also be correct by filtering techniques.

2.5 LINE DROPOUT CORRECTIONS

Lines of data which are not recorded usually manifest themselves on an image by black lines where each pixel has a DN of zero. Such line dropouts may be caused by a loss of signal as the DNs are transmitted to the receiving station or by defects in the recording of the data at the station. Landsat TM transmits data at 85 Mbits per second, so that even a momentary loss of signal can result in a line of data being lost. The visual interpretation (and statistical analysis) of an image is greatly impaired if there are many line dropouts for one scene. The lost data cannot be recovered. However, it is possible to substitute 'dummy' digital numbers for the missing ones in order to make the image more visually appealing. The commonest method for fixing line dropouts is to apply a computer algorithm to the data which assigns every pixel in the defective line a digital number which is the average of the pixel values for the adjacent lines.

$$DN_{i,j} = (DN_{i-1,j} + DN_{i+1,j})/2 \qquad \text{equation 2.1}$$

Figure 2.8a shows an MSS 7 image in which a line of data has been lost and which is represented by a black line. The result, after applying the correction shown in equation 2.1, is illustrated in Figure 2.8b. A horizontal line dropout may occur for a transverse scanning system if a detector malfunctions. However, in a pushbroom system, each pixel in any one line is observed by an individual detector. Thus a faulty detector in a pushbroom system may manifest itself by a column of pixels with a DN of zero. The underlying assumption in applying equation 2.1 is that the adjacent DNs will provide a guide to the DN that would be expected to have been recorded if the data had not been lost. The extent to which this assumption is justified will depend to a large degree on the nature of the scene and the wavelengths that are being sensed. Figure 2.9 shows the DN for 18 pixels (6 pixels by 3 lines) for 6 TM bands. The centre line

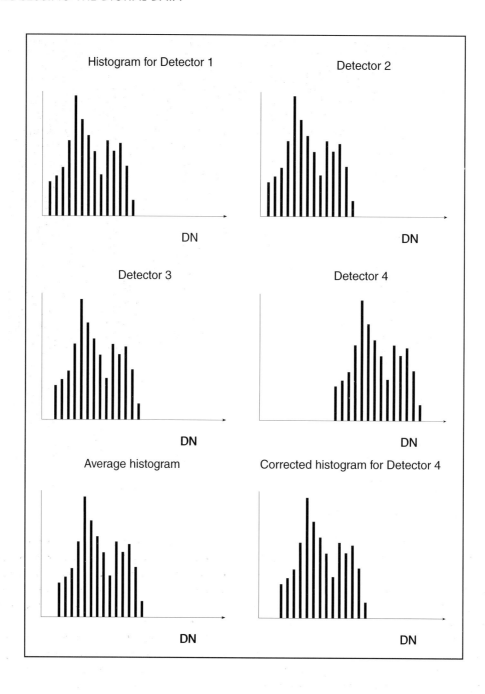

Figure 2.6 Correction to data obtained by faulty detector (4) by comparing histograms to an average histogram for all detectors.

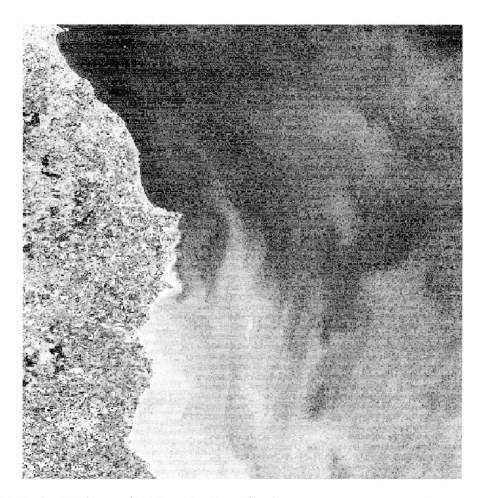

Figure 2.7 (a) Landsat MSS 4 image of Irish Sea with evidence of banding.

for each of the TM bands shows the actual recorded DN values and also the computed DN if a line dropout program is run. In general, the computed DN and the actual DN are quite close for most bands, agreeing to within 1 or 2 DN. However, the computed and actual value for TM 4 diverges to a much greater extent.

Most digital image processing systems have the necessary computer program to carry out the above procedure but a number of difficulties may arise. If the first line of data is defective, then the line is assigned the DN of the line below it. Similarly, if the last line is missing, it is assigned the same DN as the line above it. However, not all image processing systems have the capabilities to correct partial line dropouts where data for half a line are missing. Thus one must either correct the entire line and lose actual data or not perform a correction. For any given line, only certain bands might have been lost. If the latter circumstance prevails, it is important to determine that the line dropout program offers the option of choosing the bands on which it operates, otherwise real data in those bands that have been recorded will be replaced by computed data.

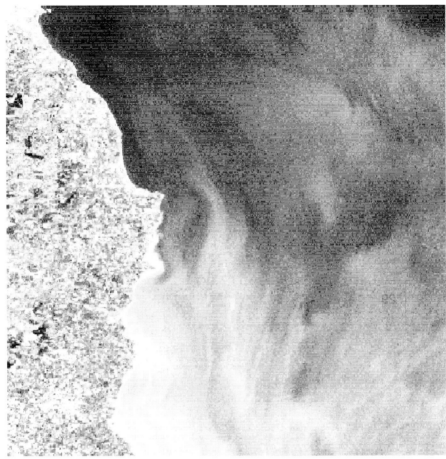

Figure 2.7 (b) Reduced banding effect after correction.

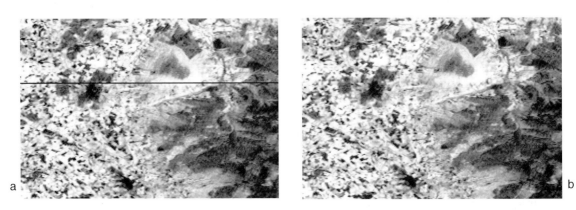

Figure 2.8 (a) Line dropout defect and (b) after correction.

Computed / Actual

TM1

74	74	74	73	75	81
73 / 74	73 / 74	74 / 74	73 / 74	74 / 74	78 / 80
73	73	75	74	74	75

TM2

32	32	31	32	33	36
31 / 32	32 / 32	32 / 33	32 / 33	32 / 32	34 / 35
30	32	33	33	32	32

TM3

28	29	29	29	31	37
26 / 25	28 / 27	28 / 29	29 / 29	29 / 29	33 / 37
25	28	28	29	28	29

TM4

122	105	88	90	87	82
123 / 132	107 / 113	98 / 94	103 / 100	98 / 93	90 / 88
125	109	108	116	109	98

TM5

77	74	72	67	75	83
71 / 71	70 / 67	72 / 67	72 / 72	75 / 77	77 / 81
66	66	72	78	75	71

TM7

22	23	24	23	28	38
21 / 19	21 / 21	23 / 22	23 / 23	25 / 26	31 / 33
20	20	22	23	23	25

Figure 2.9 Comparison of actual and computed line dropout values. Taking the average of the line above and below shows little divergence from the actual vales for most bands.

2.6 GEOMETRIC CORRECTIONS

An image produced from the digital numbers obtained by a satellite system is distorted and geometric corrections have to be applied to the data. Image distortions are generally divided into two types: systematic distortions which are predictable and generally apply to all images obtained by a specific remote sensing platform and non-systematic distortions which apply to individual images. There are a number of sources of systematic distortion, the most obvious of which is that due to the Earth's rotation. Landsat obtains its data in a southbound direction by scanning sets of lines (six lines simultaneously for MSS). However, in the time between Landsat scanning its first and last line (approximately 28 seconds for the Landsat system), the Earth has rotated eastwards beneath the satellite. Therefore for a 185 km × 185 km scene, the ground area image is not square but a rhombus shape (Figure 2.10). Other systematic distortions include velocity variations in the scanning mirror and forward movement of the satellite during the acquisition of data for a scan line (scan skew). Two other points are worth noting regarding distortions on satellite images. Pushbroom systems such as SPOT do not obtain their data by

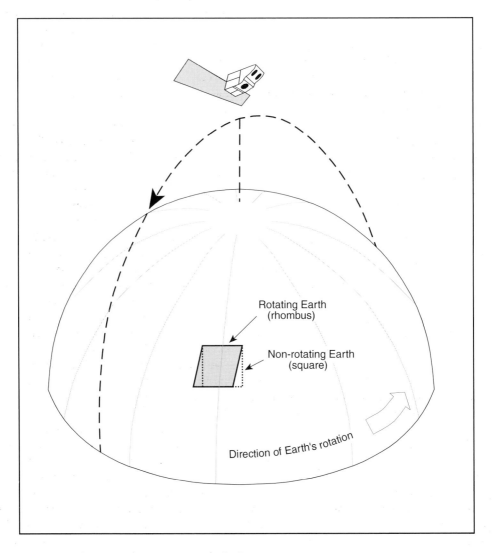

Figure 2.10 Distortion due to Earth's rotation. See text for discussion.

sweeping a strip of terrain with a scanning mirror; scanning mirror and scan skew distortions thus do not apply. Second, upon examination of an image, north appears to be at the top and the scan lines appear to run in an east–west direction. However, most Earth resource satellites are in a near-polar orbit which diverges by about 9 degrees from a true polar orbit. Most systematic distortions can be corrected using data obtained from accurate monitoring of the satellite's orbital path and a knowledge of the scanning system's characteristics. These are normally corrected before the data are delivered to the user. Non-systematic distortions result mainly from variations in the satellite's orbit. The platform is not perfectly stable and variations in the orbital height may occur. The attitude of the system may also be altered by pitching, rolling and yawing.

The procedure for geometrically rectifying an image using a digital image processing system is explained below using part of an MSS Landsat scene as an example. Figure 2.11a shows an infrared image of Lough Neagh in Northern Ireland. The land–water contact is well displayed because of the high contrast between the black signature for the water and the brighter signature for the surrounding vegetation. The dimensions of Lough Neagh, taken from a 1:250,000 map of this region, are shown in Figure 2.11b. When it is compared with Figure 2.11a, the spatial distortion of the image is obvious. The distortion is so great on the image because each pixel in an MSS scene is 56 m across track by 79 m along track. However, on an image display system, this rectangular area is represented on the screen by a square pixel. Thus one can visualise the rectangle as being 'stretched' in an east–west direction by about 41 per cent ($23/56 \times 100$) to make it square. Thus, in order to use the image to determine spatial and angular relationships, it must first be rectified. The rectification process involves a number of stages. However, in essence it involves the comparison of an uncorrected image with a georeferenced dataset. The georeferenced dataset may be a map or another image. The comparison is made by means of ground control points (GCPs). A ground control point is a point whose position can be determined on the uncorrected image (row and column position) and also on the georeferenced dataset (latitude, longitude or grid co-ordinates). The first stage involves the collection of ground control points to 'tie down' the uncorrected image (Figure 2.12a). There are a number of alternative methods of achieving this tie-down. If the uncorrected image is being tied down to a georeferenced image, both are displayed simultaneously on the screen and GCPs collected from either image. An alternative approach is to correlate the uncorrected co-ordinates with a map placed on a digitising tablet. A third approach is to collect the co-ordinates of the GCPs from a map and manually input the data into the computer. Once the ground control points are collected, the pixels in the uncorrected image are transformed to the georeferenced database by means of warping polynomials.

Figure 2.11c shows how Lough Neagh appears on a Landsat image after it has been rectified. A comparison of Figures 2.11b and 2.12c illustrates that the rectified image now more closely resembles the correct shape. Rectification is a translation that is simply a mathematical process. Whether it produces the correct results or not is dependent upon the operator understanding this translation. Figure 2.11d is a rectified version of the original uncorrected image. In this instance the mathematical procedure has operated correctly, but the results are clearly erroneous. The rectified Lough Neagh is not only upside-down, but also inside-out! This has occurred because the co-ordinate system for the georeferenced data has its origin at the lower left (thus co-ordinate values increase northwards and eastwards) whereas the origin for the uncorrected image is at the top left. Pixel co-ordinates thus increase in an eastward and southward direction.

The degree to which the rectification process is successful or otherwise also depends to a large extent on the choice of ground control points. A good ground-control point is a feature which can be clearly identified and its co-ordinates obtained on both the map (which shows the correct co-ordinates) and the unrectified image and which does not alter its position in time. For example, consider rectifying a 1998 SPOT image to a 1990 map. The spatial relationships

Polynomial Transformation

A polynomial is essentially a mathematical equation which links the uncorrected image to the georeferenced database based on the input ground-control points. The polynomial dictates the translation, scaling and/or rotation that the uncorrected image has to undergo in order to be georectified. A polynomial equation may be of any order. A first-order translation links the uncorrected pixel co-ordinates (x_u, y_u) to the georeferenced co-ordinates (x, y) by:

$$x_u = a_0 + a_1x + a_2y$$

equation 2.2

$$y_u = b_0 + b_1x + b_2y$$

A more complicated polynomial transform is a second-order one (quadratic) in which the uncorrected image and georeferenced database are related by the equation:

$$x_u = c_0 + c_1x + c_2y + c_3xy + c_4x^2 + c_5y^2$$

equation 2.3

$$y_u = d_0 + d_1x + d_2y + d_3xy + d_4x^2 + d_5y^2$$

The operator may choose the order of the transformation polynomial, though some image processing systems offer a limited choice, for example up to order 5. In theory, the higher the order, the more closely a polynomial will fit the georeferenced database. However, in practice, whilst a higher polynomial will produce a more accurate fit at localities near a ground control point, other errors may be introduced at large distances from GCPs. The computing time for a rectification process using a higher order polynomial may also be substantially greater than for a first-order, linear transformation. To produce a full first-order transform (translation, scaling and rotation), a minimum of three ground control points is required whereas 22 GCPs are needed for a full fifth-order transform. A first-order transformation may be sufficient, though in practice far more than three GCPs are used. The constants in the polynomials are derived using a least-squares regression method to minimise the errors for each GCP. The polynomial will not transform every GCP with 100 per cent accuracy and it is important before the full rectification proceeds to ensure that the errors are within acceptable limits. A root mean square (RMS) error for the GCPs can be calculated. This is a statistical measure of the error between the calculated co-ordinates of a GCP and the co-ordinates of the GCP in the georectified image. Figure 2.12b shows that all GCPs except number 6 have an error of less than one pixel in the x and y directions. The error for GCP 6 is much larger and suggests that a mistake may have been made when its co-ordinates were entered. This GCP should be re-examined and corrected or discarded and not used in the rectification process. Once the GCPs have been checked and the RMS error is within acceptable limits, the uncorrected image can be rectified to the georeferenced database.

for some features will have changed in the eight-year gap between the map and image. The geographical co-ordinates of the corner of a forest in 1990 may differ because of further planting or felling of the timber. Rivers evolve over time and continually alter course; a GCP at a prominent bend may thus no longer be accurate. Although the river may not be a suitable GCP, one end of a bridge that spans the river is a feature that would not be expected to have altered its position. Similarly, the centre of a prominent roundabout or where two airport runways intersect would be suitable ground control points. Even a 1998 map will not be contemporary with the image because of the lag time in map production. It is possible to obtain accurate positional data of selected ground control points during a satellite pass by using a global positioning system, though it seems rather ironic to be using satellites to pinpoint a position in

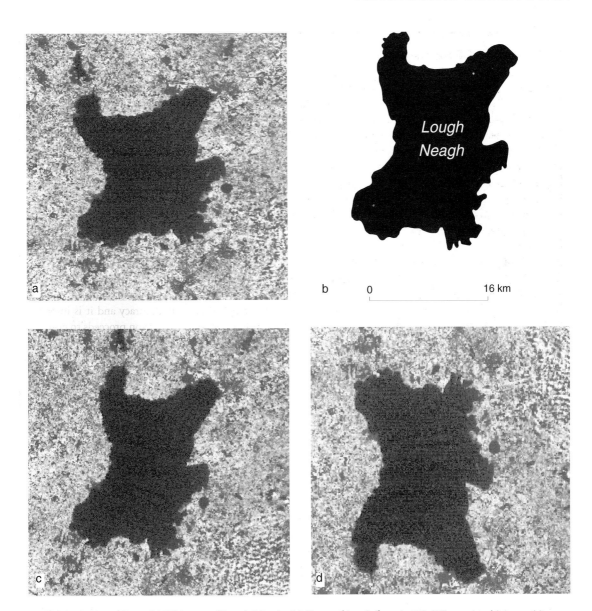

Figure 2.11 (a) Uncorrected MSS image of Lough Neagh. (b) Shape of lough from 1:250,000 topographic map. (c) Geometrically rectified MSS image of Lough Neagh. (d) Error in rectification process.

order that images produced by data from other satellites can be properly rectified.

Identifying ground control points is not a sufficient requirement to ensure a correct rectification: the number and relative position are also of great importance. Figure 2.13a illustrates a situation where a

large area is being rectified with insufficient GCPs. Although the rectification procedure would operate mathematically, the rectified image is likely to be inaccurate. Figure 2.13b shows a situation in which a greater number of ground control points have been obtained. However, they are concentrated in the

24 PRE-PROCESSING THE DIGITAL DATA

western part of the image, so that the rest of the image is not tied down sufficiently and again the rectified image may be inaccurate. If the GCPs are concentrated in two opposite corners of the image, the geometric correction would not be accurate in the remaining parts of the image. The analyst is often faced with a dilemma when attempting to georectify an image. In order to obtain sufficient ground control points whose positions encompass the entire image, he/she may be forced to use points whose positions cannot be absolutely guaranteed. In the situation shown in Figure 2.13c, using ground control points delineated by rivers or the coast (which may alter position) may well yield a more accurate georectified image than performing the rectification procedure without them. A further problem with coastal features is that the map may refer to low tide whereas the image may have been obtained when other tidal conditions prevailed.

In order to complete the entire rectification process, each pixel in the corrected image has to be assigned a new DN. At first glance this may seem a superfluous procedure but an examination of Figure 2.14 illustrates that such a procedure is required. Rectification does not produce a simple one-to-one translation in which, for example, the pixel value in row 5 and column 6 of the uncorrected image is translated to the pixel value in row 5 and column 6 of the rectified image. A pixel in the rectified image is overlaid by a number of pixels in the uncorrected image; consequently a new digital number has to be assigned to the corrected pixel. There are a number of different techniques for assigning a new digital number to a rectified pixel based on interpolation. The simplest approach is known as nearest-neighbour interpolation and the new DN is the DN of the uncorrected pixel that is closest to the co-ordinates of the corrected pixel. Thus, in the example shown in

Projections

Implicit in the above discussion has been the assumption that the unrectified image is distorted while the map is totally correct. However, it should be realised that a map is a two-dimensional representation of a three-dimensional surface – the Earth's globe. Conversion of the features on the globe to a flat map involves the concept of projections. One can envisage the projection of the continents onto a surface such as a cone, plane or cylinder surrounding a transparent globe with a light source at the centre. Many types of projection exist but it is not possible simultaneously to preserve area, distances, directions and shapes. A Mercator or Lambert Conformal Conic (conformal projections), for example, preserve shapes whereas a Sinusoidal or Albers Equal Area projection preserves area. In order to project the Earth, which is an irregular oblate spheroid, it is necessary to employ an 'Earth model', which is an ellipsoid taken to represent an idealised Earth.

The position of a place on a map is generally indicated either by latitude and longitude or on a grid-based co-ordinate system. Whilst the latitude and longitude convention is unique, an individual country may have its own grid-based co-ordinate system that applies only to that country. A typical scenario is where large areas (such as 100×100 km squares) are designated by letters and within each square each position is given a northing and easting. Such maps employ a grid north with equal north–south and east–west scales whereas the distance between two lines of longitude decreases as one moves north or south of the equator. Remotely sensed data are increasingly being used in Geographical Information Systems that often employ many layers of data. These data are generally obtained from various sources incorporating a range of co-ordinate systems. In such situations, it is necessary to convert the data to a single co-ordinate system. Most image processing systems allow this conversion between different systems. PCI software, for example, supports 20 different Earth models and 25 projections.

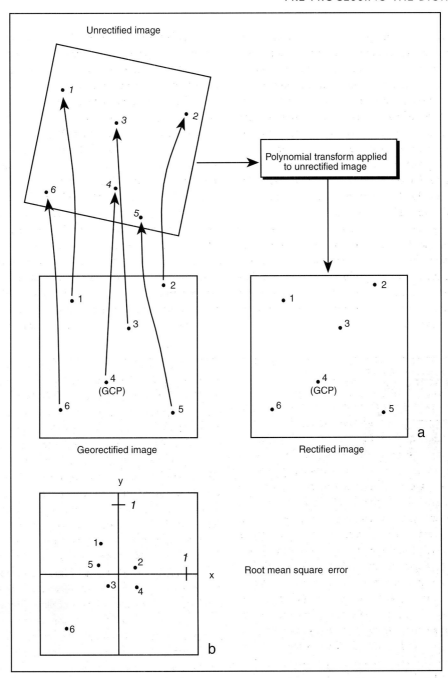

Unrectified image

Polynomial transform applied
to unrectified image

Georectified image

Rectified image

a

y

1

5

2

1

3

4

6

x

1

Root mean square error

b

Figure 2.12 (a) Rectification process. Unrectified image is tied down to a map or rectified image and the polynomials calculated. The unrectified image is then transformed using the polynomial equations. (b) Graphical expression of root mean square error. In this instance ground-control point 6 appears to be in error, possibly due to an inaccurate tie-down or the input of incorrect co-ordinates.

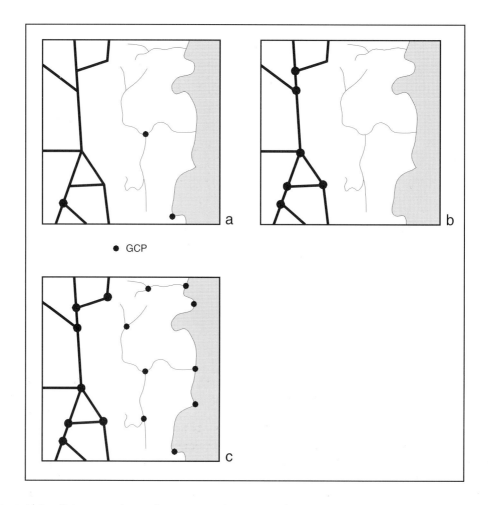

Figure 2.13 (a) Insufficient ground control points to yield a good rectification. (b) Poor distribution of ground control points to yield a good rectification. (c) Good rectification because of use of large number of evenly distributed ground control points.

Figure 2.14, the shaded pixel is assigned a value of 16. A bilinear interpolation produces a new DN that is the weighted average of the four closest pixels. With this transform the new DN for the shaded pixel would be 12, assuming each pixel is given equal weight. Computing time for a bilinear interpolation is greater than for a nearest-neighbour interpolation and image resolution is not as good because of the smoothing effect of this method. Cubic convolution employs a weighted average of the closest 16 pixels, which further increases the computing time.

2.7 ATMOSPHERIC CORRECTIONS

When a detector on a remote sensing system measures the radiance of a pixel and assigns it a DN, the DN is formed of two components. One is the actual radiance of the pixel which we wish to record, but added to it is an atmospheric component (Figure 2.15). The atmospheric component has a scattering effect which tends to increase the DN that is assigned to the pixel and also an absorption effect which may reduce the assigned DN (Jensen 1996). Scattering in

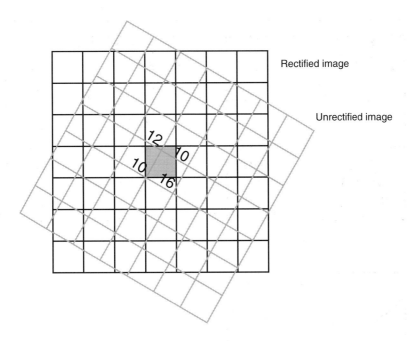

Figure 2.14 Resampling of digital numbers for rectified image.

the atmosphere can result in the pixel being indirectly illuminated by scattered radiation or in radiance being recorded by the sensor because of scattering from particles within the atmosphere. The extent to which the atmosphere alters the true DN is best seen by an examination of the DN histograms for different bands. Many scenes contain very dark pixels (such as those in deep shadow) and it might be assumed that they should have a DN of zero. However, when the histograms for different bands are examined, some are seen to be offset from zero (Figure 2.16). Scattering is inversely proportional to wavelength; thus shorter wavelength bands have a greater offset from the origin. The degree of offset is dependent on the atmospheric conditions that change laterally and temporally; it is therefore not possible to give absolute offsets which are applicable in all situations. For Landsat TM, typical offsets are of the order of TM 1: 25–35; TM 2: 20–30; TM 3: 10–20; TM 4: 5–15; TM 5: 0–5; TM 6: 0–5; TM 7: 0–5.

A first-order atmospheric correction may be applied to remotely sensed datasets by assuming the offsets are due solely to atmospheric effects and subtracting the offset from each DN. Using the average offsets given above, a DN of 70 in TM 1 would be corrected to a DN of 40 (70 − 30); a DN of 70 in TM 2 to TM 7 would be 45, 55, 62, 67, 67, 67 respectively. Atmospheric corrections are not a necessary prerequisite before remote sensing images are enhanced. False colour composites produced from data uncorrected and corrected for atmospheric effects would appear very similar to our eyes. However, the atmospheric correction is important in the construction of ratio images and as a rule it is best to eliminate this component before any further analysis. If the study is terrestrial in character and TM 1 and TM 2 are not required, then much of the atmospheric component can be eliminated by not using these bands in any further processing.

An important assumption has been made by simply subtracting the lowest DN in a dataset in order to take account of the atmospheric component. This approach takes for granted that there are pixels within the scene whose true DN is zero, which may not be the case. Consider a TM 1 image of a region covered by snow and ice. The lowest DN in this scene

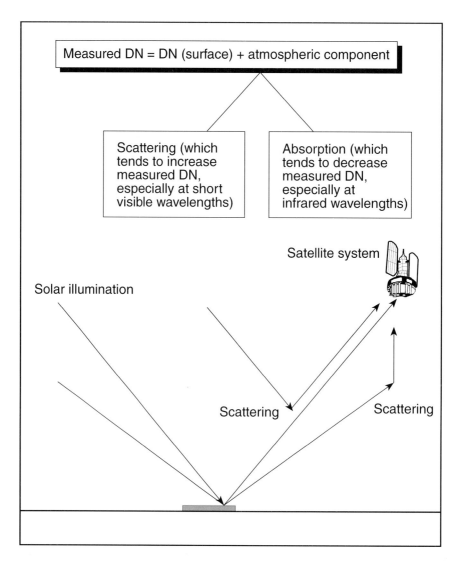

Figure 2.15 Atmospheric contribution to DN measured by remote sensing sensors.

may be only 80. To subtract this value from the dataset would not be correct, as the reflectivity of snow is very high in the visible range and subtracting 80 would significantly reduce this. Again, it must be emphasised that the subtractive method of taking account of the contribution of the atmosphere, which is a very dynamic system, produces only an estimate. A better approach would be to have some knowledge of the atmospheric conditions when the data were obtained. This knowledge is rarely available except in planned research projects where forward planning can synchronise the acquisition of information on the atmosphere during the overpass of the satellite. Absorption models are particularly important in many marine applications. It may, however, be possible to acquire meteorological data if research is being

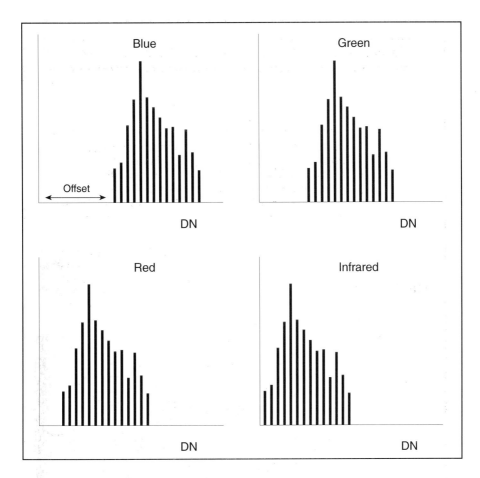

Figure 2.16 DN offsets in DN histograms resulting from atmospheric scattering. Offsets tend to be greater for lower wavelengths bands.

carried out in a country with a sophisticated meteorological observing network.

2.8 SOLAR ILLUMINATION CORRECTIONS

Images obtained at different times of the year are acquired under different illumination conditions. The solar illumination angle, as measured from the horizontal, is greater in the summer than in the winter. An important application of remote sensing data is the production of change detection images which show the changes that have taken place for a scene

between two given dates (see section 3.9). However, if two images of the same area, taken on different dates, are compared, they will not be similar even if there has been no change in the spectral characteristics of the elements within the scene because of the different illumination angles. Thus, in order to ascertain whether any changes have occurred in a region, it is necessary to remove the effects of the differing solar illumination. One method of doing this is to normalise the data by calculating for each pixel (based on the actual DN) the DN that a pixel would be expected to have at a particular illumination angle. Lillesand and Kiefer (1994) suggest that the data can be normalised to an illumination angle of 90 degrees by dividing the observed DN by the cosine of the zenith angle where the zenith angle is the angle between the solar illumination and the vertical (Figure 2.17a). An allied correction may be applied to sloped terrain. A pixel on sloped terrain will be illuminated by radiance of a different intensity than a pixel on a horizontal surface (Figure 2.17b). Thus the measured radiance for the pixels will be different even if they have similar reflectance characteristics. The most extreme situation is a pixel facing away from the illumination direction. This will receive no input energy (assuming it is not being indirectly illuminated), no energy will be reflected and it will be assigned a DN of zero. A number of topographic correction methods have been described by Civco (1989) and Teilet *et al.* (1982) which involve calculating new DNs based on each pixel's aspect and slope and which allow the pixel values to be normalised to a horizontal plane.

Another aspect of solar illumination that varies for different times of the year is the Earth–Sun distance. The Sun is at its closest to the Earth on 3 January (146.4 million kilometres) and at its farthest on 4 July (151.2 million kilometres). The reflectance of an area is dependent on the intensity of the irradiance that obeys an inverse square law. Thus, as the Sun is closer to the Earth in January (for both the northern and southern hemispheres), a pixel whose reflectance characteristics have not changed should be associated with a higher digital number on a January image than on a July image if only the Earth–Sun distance effect operated.

It is relatively easy to perform solar illumination and Earth–Sun distance corrections on a digital image processing system. Both corrections involve a simple multiplicative process which is applied to every pixel. Aspect corrections are much more complex. Such corrections would imply that the slope of every part of the scene is known to the same scale as the pixels. Thus, for a SPOT panchromatic image, values of slope would be required for 10 m $\times$ 10 m areas. Because each pixel has a different slope, a separate correction is needed for each one. Topographic corrections to satellite data can be performed if a

---

Two pixels have DNs of 50 and 200 on a July image. If their spectral properties have not changed, what digital numbers would these pixels be expected to have on a January image due solely to differences in the Earth–Sun distance?

Differences in the Earth–Sun distance would result in the pixels being brighter on the January image by a factor of $(151.2/146.4)^2$ or 1.066 because of the inverse square relationship. Therefore the pixel values in January would be 53 (50×1.066) and 213 (200×1.066). Note that, if this factor is not taken into account, the disparity in absolute terms is quite high for high values of DN but for low digital numbers the absolute difference becomes smaller.

---

digital elevation model for the scene exists or can be created.

2.9 DIGITAL DATA FORMAT

Digital satellite data may be obtained on different media from suppliers like the National Remote Sensing Centre in Britain or SPOTIMAGE in France. Initially data were supplied on computer-compatible tapes (CCT). A single CCT is termed a 'volume' and early tape formats might require more than a single tape to hold all the data for one scene for all bands. Multiple volumes for a single scene are referred to as a 'volume set'. Thus, gaining access to all the data required mounting, reading and dismounting a number of tapes. However, with the current storage capacity of CDs, an entire TM scene can be stored on a single CD from which the data can be easily transferred to the hard disk of a computer. The data format for the digital data can vary, though common formats are band sequential, band interleaved by line or band interleaved by pixel.

A separate file is used to hold each band of multi-

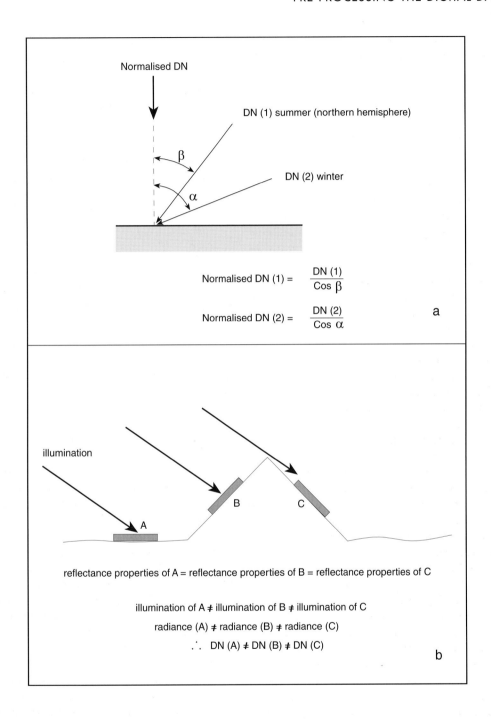

Figure 2.17 (a) Correction applied to measured DN in order to take account of different illumination angles. (b) Effect of varying aspect with respect to illumination on the measured DN.

spectral data in band-sequential format. In addition, each data file can be preceded by a header (or leader) file and a trailer file. Because the band-sequential format contains the relevant data for a single band in one file, the data for a large area for a single band can be relatively quickly retrieved. However, to access a small area but in three separate bands would require a longer processing time to subsample data from individual files.

A band interleaved by line format for an n-band image has the data for all bands for each line stored in sequence. Thus the sequence runs as follows: line 1, band 1; line 1, band 2 ... line 1 band n followed by line 2, band 1; line 2, band 2 ... line 2, band n, etc. A band interleaved by pixel format is similar except that the data for all bands for each pixel are stored thus: pixel (1,1), band 1; pixel (1,1), band 2 ... pixel (1,1), band n followed by pixel (1,2), band 1; pixel (1,2), band 2 ... pixel (1,2), band n. A typical band interleaved by line multispectral SPOT scene provided by SPOTIMAGE on CD has the following file parameters (Table 2.1):

Table 2.1 Typical band interleaved by line format for a multispectral SPOT scene obtained on CD from SPOTIMAGE

	Record length (bytes)	Number of records
Volume directory	360	5
Leader file	3,960	27
Imagery file	5,400	9,013
Trailer file	1,080	3
Null volume directory	360	1

The imagery file for a 60 × 60 km multispectral (three-band) SPOT image is approximately 48 Mbytes (5,400 × 9,013). The other files contain information on the histogram characteristics and scene parameters such as date, time, instrument, processing level, spectral mode, number of lines, number of pixels per line, orientation angle, azimuth and elevation sun angles and the scene centre location.

2.10 CHAPTER SUMMARY

- Before the data obtained by a remote sensing system can be analysed, certain corrections are applied. These corrections are necessary because of detector defects, loss of data during transmission and the effects of the atmosphere.

- A DN histogram is a graphical representation of the DN distribution for a scene. Some histograms may approximate to a normal distribution whereas others may be skewed. It is possible to calculate some statistical parameters for the histograms such as mode, mean, median and standard deviation.

- To view a false colour composite, it is important that the input bands are correctly co-registered with each other.

- Line-banding corrections are necessary because the response of detectors on a remote sensing system may not be constant. This may be manifest on an image by a banding effect in which every nth line is consistently paler or darker than the other lines.

- Lines (or partial lines) of data lost due to transmission errors may be replaced by 'dummy' data. A commonly used algorithm in this process substitutes data in the lost line by data obtained from adjacent lines or by an average pixel value derived from the two immediately adjacent lines.

- An image obtained by a remote sensing satellite may be distorted by, for example, the Earth's rotation and the satellite's orbital path. Rectification is a process which minimises these distortions. A commonly used rectification procedure involving ground control points links co-ordinates of features in the distorted image with the co-ordinates of the same features on a map which is spatially correct.

- The digital number obtained by a sensor may incorporate an atmospheric component as a result of scattering. Short-wavelength bands are affected to a greater degree than longer wavelength ones. A first-order atmospheric correction can be applied to the data by subtracting the lowest DN (in each band) from every DN in that band.

- Corrections may be applied to a dataset to take

account of varying solar illumination effects. The illumination angle and the Earth–Sun distance vary between summer and winter.

- Common digital data formats are band sequential, band interleaved by line and band interleaved by pixel. In band-sequential format, the data for each band are held in a separate file. For band interleaved by line and pixel, data for different bands are stored on the same file.

SELF-ASSESSMENT TEST

1 If a skip factor of 4 is used in the construction of a DN histogram, what percentage of DNs is used?

2 What is a line dropout and how does it differ from line banding?

3 What are the characteristics of a good ground-control point?

4 How does the Sun's illumination angle vary between winter and summer in the northern hemisphere?

5 How are polynomials used in geometric correction?

FURTHER READING

Civco, C. L. (1989) 'Topographic normalization of Landsat Thematic Mapper digital data', *Photogrammetric Engineering and Remote Sensing* 55, 9: 1303–9.

Cracknell, A. P. and Hayes, L. W. B. (1991) *Introduction to Remote Sensing*, London: Taylor and Francis (Chapter 8).

Jensen, J. R. (1996) *Introductory Digital Image Processing: a remote sensing perspective*, Englewood Cliffs, NJ: Prentice-Hall (Chapter 6).

Lillesand, T. M. and Kiefer, R. W. (1994) *Remote Sensing and Image Interpretation*, New York: John Wiley and Sons (Chapter 7).

Mather, P. M. (1987) *Computer Processing of Remotely-Sensed Images*, Chichester: John Wiley and Sons (Chapter 4).

Teillet, P. M., Guindon, B. and Goodenough, D. G. (1982) 'On the slope–aspect correction of multispectral scanner data', *Canadian Journal of Remote Sensing* 8, 2: 84–106.

3

ENHANCING THE IMAGERY

3.1 INTRODUCTION

Once the corrections discussed in Chapter 2 have been made, the images can be displayed and enhanced. There is no single enhancement procedure that can be considered the optimum one. The best one is the one that best displays the features that are of greatest interest to the analyst. Consequently, a range of enhancement procedures has been developed that can be used in different research fields. Thus a simple contrast stretch may be employed to obtain an overall view of the appearance of an image (section 3.2) but a geologist interested in faults may filter the data (section 3.5), whereas a botanist may produce a ratio image (section 3.4). The processing is usually performed interactively; the enhancements are thus continually modified to optimise the information output in the light of the previous procedures applied to the data and a knowledge of their effects as displayed on the computer monitor. Although the enhancements are discussed individually, it is important to realise that the final image may be the result of a number of separate enhancements. The enhancement procedures discussed here include:

- contrast stretching
- ratio images
- thresholding
- density slicing
- filtering techniques
- principal component images
- Intensity, Hue, Saturation images
- synergistic images
- non-image datasets
- classification.

3.2 CONTRAST STRETCHING

Figure 3.1a shows a TM 5 image produced using digital data after appropriate pre-processing to correct any defects. The image is very dark and virtually nothing of interest can be discerned. Such a situation is typical for a large majority of remotely sensed images. The reason for such a low contrast is that many natural features within the landscape have a low range of reflectances in a specific waveband. Sensors onboard remote sensing satellites such as Landsat are also designed to record the reflectance of any surface that they may image. Thus an 8-bit detector which can record 256 grey levels must be able to record a reflectance value for the brightest snow (say 255) and also for the darkest rock (say 0). If they were not designed in this way, information about these types of surface might be lost. However, an average scene does not encompass this entire dynamic range of reflectances with DNs ranging from 0 to 255 and the DNs are often compressed into a small part of the available range (Figure 3.2a). In order to view an image, it is often necessary to 'stretch' the data. Various types of stretch are available and can be considered under the headings of linear and non-linear stretches.

Linear Stretch

The image in Figure 3.1a has digital numbers that range from 0 to 94. A linear contrast stretch assigns new digital numbers to an output image by assigning to the lowest and highest DN in the input image values of 0 and 255 respectively in the output image and stretching all intervening digital numbers accordingly (Figure 3.2b). An examination of the linear-stretched image (Figure 3.1b) allows much more information to be extracted from it than from the original unstretched image. The scene encompasses large tracts of heath-covered bog in northwest Ireland, shown by pale tones, with the Atlantic Ocean depicted by the dark signature in the north. It is important to realise that the number of 'bins' (grey levels) in the stretched image is the same as that in the input image but the distance between them has increased.

It is possible to produce a simple formula based on the above calculations that may be used to determine the digital numbers in the stretched image. Let DN_{min} and DN_{max} be the minimum and maximum DNs in the original image. Then the DN in the stretched image (DN_{st}) for any input DN is given by:

$$DN_{st} = 255 \times \frac{(DN - DN_{min})}{(DN_{max} - DN_{min})} \qquad \text{equation 3.1}$$

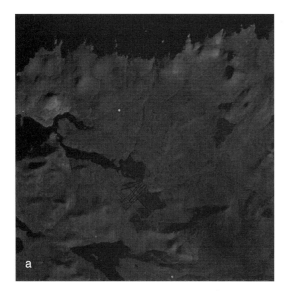

Figure 3.1 (a) Unstretched Landsat TM 5 image of Co. Mayo, Ireland, that provides very little information to the human eye. (b) Linear stretched and (c) histogram equalisation stretched versions of TM 5 image which allow a much greater amount of information to be obtained. Data courtesy of ERAMaptec.

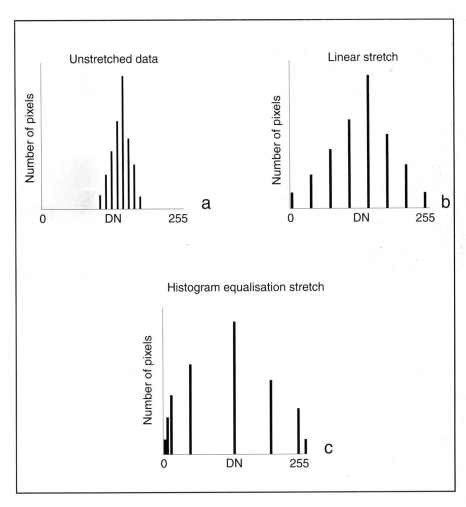

Figure 3.2 (a) Unstretched DN histogram that uses only part of the available range. Effect of applying (b) a linear contrast stretch and (c) a histogram equalisation stretch to the original data.

A histogram for a non-enhanced image is shown in Figure 3.3. Calculate the output DNs in a contrast-stretched image for the input values 60, 80 and 200.

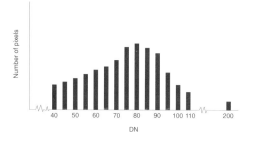

Figure 3.3 Simplified frequency histogram for use with in-text question.

The input range of 160 (200 − 40) is mapped to an output range of 0–255. Therefore each input unit is equivalent to 255/160 output units. The value 40 (lowest) in the input range is assigned a DN of zero (lowest) in the output, therefore:

60 in the input is equal to
$$20 \times (255/160) = 32$$

80 in the input is equal to
$$40 \times (255/160) = 64$$

200 in the input is equal to
$$160 \times (255/160) = 255.$$

Why do you think this linear stretch would be inefficient for the histogram shown in Figure 3.3?

In the example shown in Figure 3.3, there are no digital numbers between 110 and 200, therefore a large number of the grey levels in the output image are redundant. A more efficient stretch would be one that mapped the range 40–110 to 0–255.

Such a stretch increases the gap between most grey levels in the output image but will now assign all input DN from 110 upwards a value of 255 in the stretched image. Thus one loses a small amount of information at the bright end of the DN range (a DN of 110 cannot now be distinguished from a DN of 200) but allows better discrimination in the rest of the image. An equation similar to equation 3.1 can be produced for this situation. Let DN_{min} be the minimum DN in the original image and DN_m be the DN to be assigned to saturation (110 in the above example). Then the DN in the stretched image (DN_{st}) for any input DN is given by:

$$DN_{st} = 255 \times \frac{(DN - DN_{min})}{(DN_m - DN_{min})} \qquad \text{equation 3.2}$$

What DN will 60 be assigned if this stretch is used?

Substituting from equation 3.2 gives:

$$DN_{st} = 255 \times \frac{(60 - 40)}{(110 - 40)} = 73$$

Using this stretch, the gap between DNs of 40 and 60 in the original image is now 73 whereas the gap was only 32 when equation 3.1 was applied.

Most image processing systems perform a simple linear contrast stretch and also allow the operator to input a user-defined range to be stretched. Depending on the particular system it may be possible to stretch selectively the DN range that represents a specific number of standard deviations from the mean. Such a stretch decreases the possibility that a very high or low number, which might represent only one pixel, will be included in the stretch.

Non-linear Stretches

Linear stretches stretch equal DN ranges by equal amounts. Thus, for Figure 3.1b, a DN range of 0–30, which represents approximately 33 per cent of the input range, is allocated 33 per cent of the output range (0–85). Similarly, 31–62 and 63–94 will each be represented by about 33 per cent of the output image.

However, a linear stretch takes no account of the number of pixels in each DN bin. Over 85 per cent of the pixels are within the 0–30 range in the input image but they are compressed into only 33 per cent of the output. A histogram equalisation stretch, however, stretches the input data in proportion to the population of the DN bins and thus provides a better contrast over the most populated part of the scene. DN bins with few pixels, usually those at the extremities of the histogram, are compressed closer together (Figure 3.2c). A histogram equalisation-stretched version of the image is shown in Figure 3.1c.

Other non-linear stretches can be applied to the data that preferentially stretch some parts of the histogram compared to other parts. Consider applying a power-law stretch (x^3) and a logarithmic (base 10) stretch to a small DN range where the DN values are low (5–10) and to the same DN range when the DNs are high (220–225).

The difference for the log stretch is greater for the low digital numbers (5 and 10) than for the high digital numbers (220 and 225); therefore a log stretch preferentially stretches the dark parts of the scene (Table 3.1). The power-law stretch has the opposite effect: it will preferentially stretch the brighter parts of the scene.

If the input histogram is markedly bimodal (Figure 3.4), a better strategy might be to apply a piecewise stretch which will stretch both the populations, possibly by different amounts. Thus the

input values may be stretched but the range between the two DN populations where there are no pixels (x–y) may be compressed in the output image (x′–y′).

Look-up Tables (LUT)

A computer can use equation 3.1 to calculate the digital numbers in a stretched image. However, if that approach is adopted, it may take too long for it to be practical to produce the stretched image interactively. A 60×60 km multispectral SPOT image would require new DNs to be calculated for 27 million pixels. Assuming 500,000 arithmetic operations a second, this procedure would take nearly four minutes. However, this procedure can be speeded up greatly by the use of look-up tables. A look-up table shows the correlation between input DN and an output DN. For an 8-bit system there are a maximum of 256 (0–255) possible input and output values. For every input value, the computer calculates the corresponding output DN (Figure 3.5a). The computer then simply reads the DN for every pixel and substitutes the corresponding output DN to produce the stretched image. A graphical representation of a LUT is shown in Figure 3.5b and it illustrates a number of points.

1 The sloped part of the graph does not go through the origin. The intercept on the horizontal (input) axis, A, is due to atmospheric scattering, assuming that the original scene contained areas which should have registered a DN of zero in the absence of an atmosphere.
2 All input values greater than B are assigned a DN of 255 and thus saturated to white on the output image.
3 The gradient of the sloped portion of the graph is greater than 1. Thus a given input range will be less than its corresponding output range, which increases the contrast in the output image.

Figure 3.5c shows the LUT for the stretch in Figure 3.4. Note that the gradients for the DN ranges encompassing the DN populations are different

Table 3.1 Effects of logarithmic and power-law stretches on low and high DN values

DN	log DN	DN^3
5	0.699	125
10	1.0	1,000
Difference	**0.301**	**875**
220	2.342	1.06×10^7
225	2.352	1.14×10^7
Difference	**0.01**	**7.4 $\times$ 10⁵**

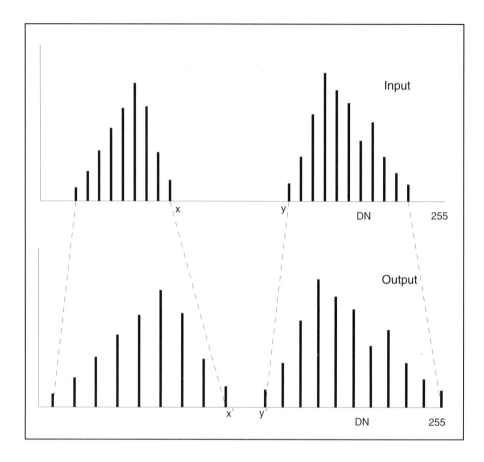

Figure 3.4 Bimodal DN histogram and piecewise stretch which stretches different parts of the original histogram by different amounts. Note how x–y has been compressed to x′–y′.

and also greater than 1, whereas the input range (x–y) where there are no pixels has a gradient less than 1.

3.3 DENSITY SLICING

It is possible to introduce colour to a single-band satellite image even though normally such an image would be displayed in shades of grey. A digital image processing system allows an operator to assign a colour to every DN in the image, though in practice a colour tends to be applied to a range of digital numbers. Such a process is termed 'density slicing'. The human visual system is more adept at discerning colours than distinguishing shades of grey; a density-sliced image is thus particularly useful for delineating different surfaces. A drawback of density-sliced images is that subtle detail is lost because a range of DNs is assigned a single colour. The number of slices and the DN range for each slice are determined interactively and are dependent on the particular scene and the features that are of interest to the operator. A single-band grey-scale image is displayed in Plate 3.1a and a density-sliced version in Plate 3.1b. Colours are assigned as shown in Table 3.2.

The density-sliced image shows much more clearly than the black and white image that the lowest DNs (blue) are concentrated in the southeast

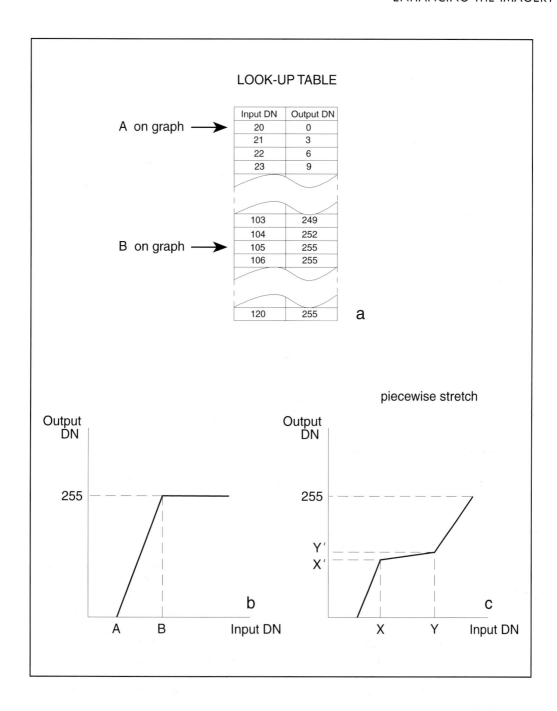

Figure 3.5 (a) Look-up table for determining output DN for every input DN. (b) Graphical representation of look-up table shown in (a). (c) Look-up table for piecewise stretch shown in Figure 3.4.

Table 3.2 Colour key for Plate 3.1b

Colour	DN range
Blue	0–75
Pale blue	76–83
Green	84–102
Yellow	103–116
Orange	117–124
Red	125–136
Pink	137–225

section of the image and there is a larger concentration of high DNs (pink) in the northwestern part of the scene. In this image many of the low DNs represent water or bare soil, whilst the bright pixels are cultivated fields. Density slicing is particularly effective where the DN values in an image represent a continuous scale for some parameter such as temperature in a water body.

A technique allied to density slicing is that of thresholding. A threshold can be applied to an image in order to isolate a particular feature that is represented by a specific range of DNs. A TM 4 Landsat image is illustrated in Figure 3.6a. When the area of the lakes in this image is being calculated, the digital

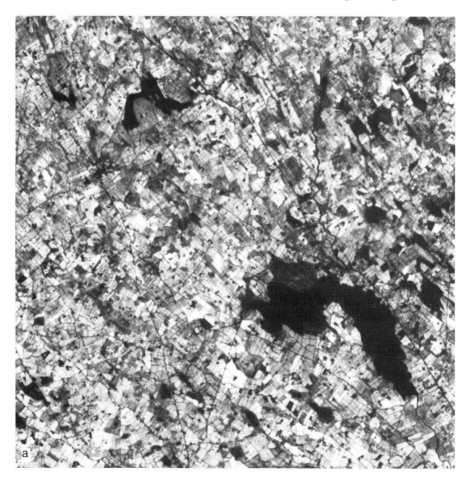

Figure 3.6 (a) Linear-stretched TM 4 image.

numbers not representing water are a distraction. The range of digital numbers associated with water here are first obtained interactively by means of a cursor, which allows the dataset to be interrogated. The highest DN for water is 35; thus this value is used as the threshold. All DNs higher than 35 are assigned a value of 255 (saturated to white) while those equal to or less than 35 are assigned a DN of zero and thus appear black (Figure 3.6b). The lakes are much more prominent in the image after thresholding. The above procedure assigns all DNs of 35 or less to zero, not just those associated with the lakes. Thus, if this image included rugged topography, then shadows which have low digital numbers might be misinterpreted as lakes.

3.4 RATIO IMAGES

The production of a ratio image, formed by dividing the digital number in one band by the corresponding DN of another band for every pixel, is one of the commonest arithmetic operations performed on digital remote sensing data (Figure 3.7a). If an area of grassland on either side of a hill is imaged in two bands, it will not have a uniform signature on either image because the digital numbers will be consistently lower for both bands in shadow (Figure 3.7b). A computer classification (see section 3.10) of this scene may correspondingly assign the sunlit and shadowed sections of the grassland to different classes even

b

Figure 3.6 (b) Thresholded version such that the lakes in the region are clearly defined.

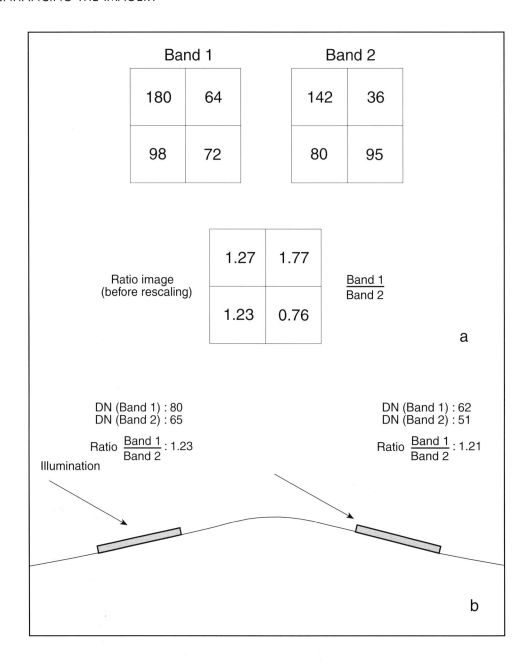

Figure 3.7 (a) Formation of a ratio image. (b) The DN for a surface can vary depending on its aspect with respect to the illumination direction. Thus in band 1, the surface has a DN of 80 on one side of the hill whereas its DN would be only 62 if it was on the other side of the hill. Similarly for band 2. However, on a ratio image, the surface has a similar DN irrespective of its aspect.

Table 3.3 Infrared and red bands for various remote sensing systems

System	Red band number	Infrared band number
Landsat MSS (1–3)	5	7
Landsat MSS (4–5)	2	4
Landsat TM	3	4
SPOT	2	3
JERS	2	3
NOAA	1	2

though they should be considered as belonging to a single class. Ratioing effectively suppresses the 'topographic albedo effect' but enhances gradient changes in the spectral reflectivity curves of different materials, which may accentuate the differences between them. The grassland example was chosen deliberately

because ratioing is particularly relevant to vegetation studies. Ratioing, to be most effective, is performed between bands for which there is a marked change in reflectance properties. Vegetation has a very high reflectance within the photographic infrared but very low in the red range. Thus an infrared/red ratio image would show high densities of vegetation as bright areas whereas bare soil, for example, would be much darker. The infrared/red band combinations for different systems are given in Table 3.3.

TM images obtained in the red (band 3) and the infrared (band 4) are shown in Figures 3.8a and 3.8b for a vegetated scene in mid-spring. The field boundaries are much more clearly defined on TM 4 and an examination of the histograms shows a much greater DN range for most of the pixels compared with TM 3. The most obvious spectral difference is the dark signatures on TM 3 (west of the circle) for some larger

Figure 3.8　(a) Landsat TM 3 image in east Ireland.

Figure 3.8 (b) TM 4 image.

fields which have a low reflectance, though on TM 4 the same fields are very bright. The ratio image (Figure 3.8c) at first glance appears similar to TM 4. Note, though, how two adjacent dark signatures in the centre of the outlined circle cannot be differentiated on TM 4 but they are resolved into a number of separate fields with different ratio values on Figure 3.8c. The ratio of infrared/red that equals 1 is referred to as the 'soil line'. Vegetated areas have ratios greater than 1. A number of other ratios referred to as 'vegetation indices' are discussed in Chapter 4. The greater the number of input bands, the greater the number of ratios that can be formed. SPOT 1, 2 and 3 have three input bands and three ratios can also be produced (1/2, 2/3 and 1/3). This assumes that the ratio 1/2 is the same as 2/1. A reciprocal image merely inverts the tones of the original ratio and provides no extra information. The six reflective TM bands (twice those

of SPOT) allow 15 ratios to be produced (five times those of SPOT). False colour composite ratio images can be constructed by projecting different ratios in red, green and blue. A simple formula allows the number of colour ratio combinations for Landsat TM to be calculated easily.

$$\frac{n!}{r!\,(n-r)!} \qquad \text{equation 3.3}$$

For this equation, n is the number of bands (15) and r is 3, as 3 ratios are needed for the false colour image. The symbol (!) refers to factorial and n! is (n) multiplied by (n − 1) multiplied by (n − 2) … 1. Thus, from equation 3.3, there are 455 different colour ratio combinations $(15 \times 14 \times 13)/(3 \times 2 \times 1)$ that can be formed by using the six reflective TM bands. In addition, many other ratios can be produced by using

Figure 3.8   (c) Simple vegetation index using TM 4/TM 3. Width approximately 5 km.

simple arithmetic calculations. An infrared/visible ratio image can be produced by dividing (TM 4 + TM 5 + TM 7)/(TM 1 + TM 2 + TM 3). In order to use ratios efficiently, a knowledge of the reflectance characteristics of different features is required.

Ratio images are often employed in the search for unknown mineral deposits. The Andes mountains in South America are located at a destructive plate margin which is associated with a number of important porphyry copper deposits. Part of the western cordillera in Peru is shown in a standard false colour composite produced by projecting TM 5 in red, TM 4 in green and TM 3 in blue (Plate 3.2a). The colours on this false colour composite are bland and the most prominent features are the many ridges, especially in the northeast, which are over 2,000 m high. Although a standard false colour composite uses three input bands, it is possible to employ all six reflective TM

bands on a ratio image. The ratio image shown in Plate 3.2b was produced by projecting bands 1/2 in blue, 3/4 in green and 5/7 in red. The topography is less prominent but colour saturation is much more pronounced, the purple upland area to the northeast contrasting well with the green signature of the lowlands in the southwest. An economic porphyry ore body, elliptical in shape and associated with a unique orange colour, can be seen in the northwest of the scene. The presence of this ore body is not evident in the standard false colour composite. An examination of the rest of the image shows the presence of isolated orange signatures in the southeast quadrant. Field investigations have shown that these regions contain ore deposits that are at present uneconomic to exploit.

Although ratio images have many advantages, it is worth considering the ratio process in some detail in order to illustrate some of the complexities that are involved. A simple ratio between two Landsat TM bands for four pixels is shown in Figure 3.9. Ratioing two numbers typically yields a real number (one with numbers after the decimal point). Discounting division by zero, the ratio may be as high as 255 but in practice the ratio is quite low, generally less than 5. Before a ratio image can be displayed it is scaled in order to increase the dynamic range and converted to a whole number (one without a decimal component). It would not be justifiable simply to perform the ratio procedure shown in Figure 3.9a. As discussed in section 2.7, the measured DN has an atmospheric component which is different for each band. In the example shown, assume the atmospheric correction for TM 4 is 0 and for TM 1 16. This component must be subtracted from the measured DNs before the ratioing is attempted. Thus the true ratio for pixel (1,1) is not 1.5 but 7.5, a five-fold increase. A zero in

the numerator with a real number in the denominator gives a ratio of zero. However, division by zero yields a value of infinity, which is defaulted to 255, though the operator may have the option of setting this value to zero as well. Division of zero by zero is not defined mathematically and the ratio may be assigned a ratio value of zero. Division by zero or having a zero in the numerator can be prevented by adding a DN of 1 to every pixel in both bands after the atmospheric corrections have been made. For high digital numbers this approach may be justified because the ratio of 223/145 will be very close to 224/146. However, at low digital numbers this process may produce significant errors: a ratio of 10/1 is not almost the same as a ratio of 11/2. If a normalised ratio is produced, where the difference between two bands is divided by the addition of the same two bands, the ratio will vary between −1 and

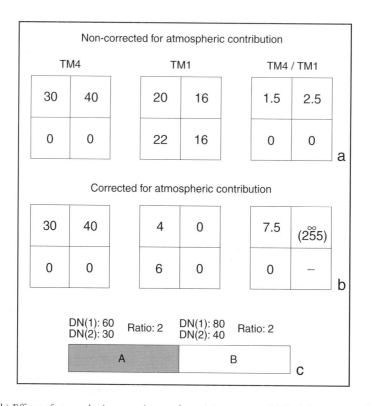

Figure 3.9 (a) and (b) Effects of atmospheric correction on the ratioing process. (c) Ratioing may result in two surfaces that could be distinguished on two separate bands being indistinguishable on a ratio image.

+1 (assuming division by zero equals zero). Before such a range can be displayed, a constant must be added to every DN to bring them all into the positive range before the application of a scaling factor.

Although ratio images can often provide information that cannot be obtained from the bands singly, there may be occasions when a ratio image provides less information. Two surfaces (A and B, Figure 3.9c) can be easily differentiated on either band 1 or band 2. In both instances surface B has a higher DN and will consequently appear paler. However, both surfaces have the same ratio value and would be indistinguishable on a ratio image of bands 1 and 2.

3.5 FILTERING TECHNIQUES

If a vertical or horizontal section is taken across a digital image and the DNs plotted against distance, a complex curve is produced. An examination of this curve would show sections where the gradients are low, correspond with smooth tonal variations on the image and sections where the gradients are high, are where the digital number values change by large amounts over short distances. Any complex curve (Figure 3.10) can be formed by the summation of a series of regular sinusoidal waves with different wavelengths and amplitudes. The curve in Figure 3.10 can be formed by combining long-, medium- and short-wavelength waves. A long-wavelength feature may also be termed a low-frequency one and a short-wavelength one is also known as a high-frequency one. The operator may wish to examine a feature or features that have a specific wavelength. The information being provided by other wavelength features can form unwanted distractions. Filtering is a process that selectively enhances or suppresses particular wavelengths within an image. Filtering is different from ratio imagery in a number of ways. Ratioing is generally applied to the entire image and a new ratio image is produced. Although an entirely new filtered image can be produced, it is possible by means of masks to confine a particular filtering procedure to a specific part of the image. Thus, for example, an urban planner may isolate an urban area in the image

in order to apply a filter which may not be appropriate for other parts of the scene but which has been specifically designed to enhance those features of interest in the urban landscape. A ratio is performed on a pixel-by-pixel basis and the DNs in adjacent pixels do not affect the calculations. However, the values of adjacent DNs are of great significance in filtering. A minimum of two bands is needed to create a ratio image whereas filtering is often performed on a single band. There are two main approaches to digitally filtering data: convolution filtering in the spatial domain and Fourier analysis in the frequency domain.

Convolution Filtering

A filter is a regular array or matrix of numbers which, using simple arithmetic operations, allows the formation of a new image by assigning new pixel values depending on the results of the arithmetic operations. The technique is illustrated in Figure 3.11. The convolution matrix in this example is a 3×3 square though rectangular matrices can also be used but they must contain an odd number of rows and columns. The matrix is initially positioned in the top left corner of the image and a new digital number calculated for the pixel covered by the central matrix cell (row 2, column 2), as shown in Figure 3.11. The matrix is then passed through the image data set by shifting one column and the same calculation is performed and a new DN applied to the pixel in row 2, column 3. In the example shown, the equal weighting of unity for the entire matrix has the effect of lessening the DN variation in the original image and smoothes the image, producing a blurring effect. Such an averaging filter reduces the effects of the noise component of an image and is useful if one is interested in the gross tonal variations rather than the fine detail. Figure 3.12b shows the effect of a 3×3 mean filter applied to Figure 3.12a. The linear ridges in the original image are less prominent in the filtered image. Figure 3.11 and the example shown in Figure 3.12b both employed a 3×3 matrix. However, it is possible to use larger arrays that, for a low-pass filter, increase the blurring effect. An 11×11 mean filter was employed to produce the image shown in Figure 3.12c.

As shown in Figure 3.11, the application of a 3×3

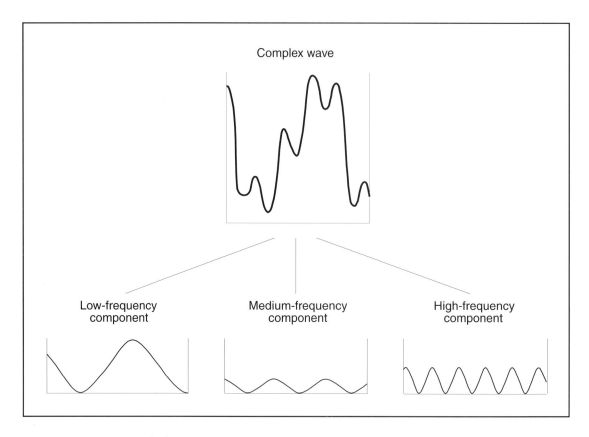

Figure 3.10 Complex curve broken down into three sinusoidal waves of low, medium and high frequency.

filter will produce an image that is two rows and two columns smaller than the original image. Larger filters decrease the size of the filtered image by greater amounts. Filtering programs are usually designed to yield filtered images that are the same size as the input image. This is achieved by replicating and extending the edge DN outside the boundaries of the original image in order that the filtering algorithm can be applied to all input pixels.

Applying high-pass filters emphasises the edges within an image. An edge on an image is represented by an abrupt change in tone (DN). Edges are generally formed by long linear features such as ridges, rivers or hedges. However, urban landscapes also contain many edges formed by roads, streets, railways and the outlines of buildings. The following section illustrates the effects of various filters upon the edges

in Figure 3.12a. Edges are important in geological investigations because they often show the position of faults and fractures which in turn can control the location of important mineral deposits.

Various types of high-pass filters exist: one of the simplest consists of a high positive number for the central DN surrounded by low negative numbers (Figure 3.13). The effects of this filter are shown in Figure 3.12d. Although this process clearly delineates the high-pass features in an image, it is quite difficult to interpret in the absence of the low-frequency features. However, if Figures 3.12a and 3.12d are combined arithmetically, the resultant image is similar to the original but appears sharper because of the edge enhancement (Figure 3.12e). (Combining the two images increases the dynamic range, so that the resultant image has to be rescaled to 0–255). A number of

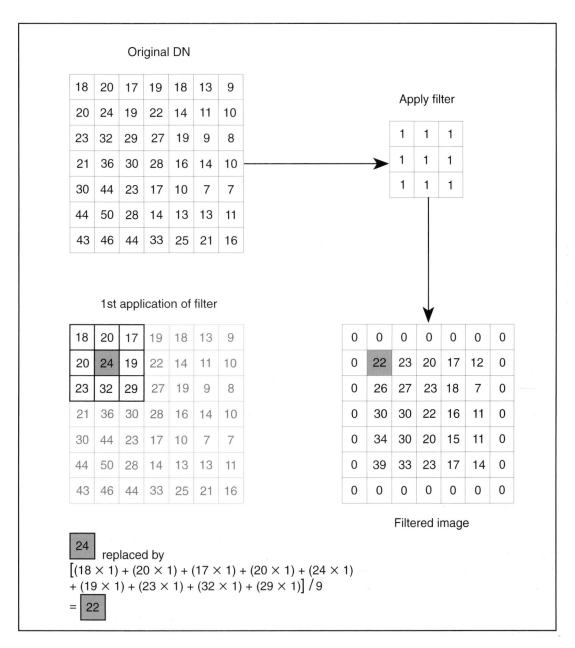

Figure 3.11 Convolution filtering process. 'Dummy' values are inserted round the perimeter of the filtered image, usually by using replicated values.

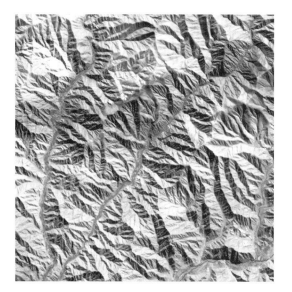

Figure 3.12 (a)

Figure 3.12 (c)

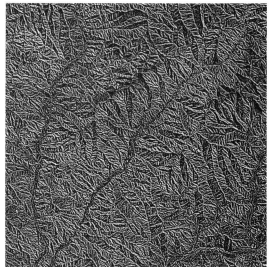

Figure 3.12 (d)

filter designs can be created which accentuate edges with preferred orientations while suppressing edges with other trends. This may prove particularly useful in a geological investigation where one wished to view features with particular orientations (which may be associated with mineralised bodies) without being dis-

tracted by other trends. First-derivative or gradient filters can accomplish this (Figure 3.13). Jensen (1996) gives examples of 30 filters, one of which he refers to as an embossing filter (Figure 3.13). When the high-frequency information provided by this filter (Figure 3.12f) is added back to the original, it pro-

Figure 3.12 (e)

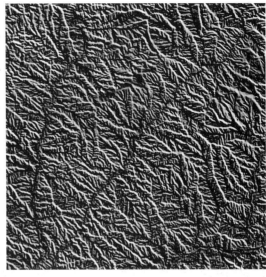

Figure 3.12 (f)

Figure 3.12 (g)

Figure 3.12 (h)

duces an image which appears sharper than the original and also has a suggestion of a three-dimensional effect (Figure 3.12g). A combination of filters can be introduced to the data. The Prewitt and Sobel filters (Figure 3.13) employ two arrays in order to detect the edges within an image. An example of the effects of

the Sobel filter is shown in Figure 3.12h. When the Prewitt filter was applied to the same image it yielded an almost identical image. An arithmetic combination of the Sobel and original image, shown in Figure 3.12i, tends to sharpen up edges irrespective of their trend. From the preceding discussion it is clear that

Figure 3.12 (i)

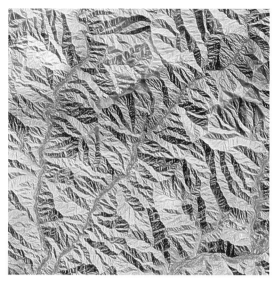

Figure 3.12 (j)

Figure 3.12 Effects of filtering. (a) Original image; (b) 3 × 3 mean filtered image; (c) 11 × 11 mean filtered image; (d) high-pass filtered image; (e) combination of high-pass and original image; (f) NW embossing filter; (g) combination of embossed and original image; (h) Sobel filtered image; (i) combination of Sobel and original image; (j) Laplacian filtered image. Data courtesy of Neil Quarmby of IS Limited.

the operator has a number of alternatives to choose from. Most image processing systems have a number of preset filters which can be employed and their effects observed on the display screen before a decision is made about whether to retain the filtered image. PCI software, for example, has a 'gamma' filter specifically designed to remove noise from radar images while retaining the edge information. A program that allows image analysts to construct their own filters is also usually available. Ultimately, the optimum filter is the one which best displays the features which are of most interest to the image analyst. In the authors' opinion, for this particular scene, the second-derivative embossing filter and the Laplacian filter (Figure 3.13) suggested by Jensen (1996) produced the best images (Figures 3.12g and 3.12j).

Frequency Filtering

Whilst the application of convolution filters permits filtering in the spatial domain, it is possible to per-

form filtering by using Fourier analysis techniques, which yield similar results, in the frequency domain. Applying the Fourier transform to the image makes it possible to break it down into separate frequencies. The mathematics involved are beyond the scope of this book but detailed discussions are provided by Teuber (1993) and Morrison (1994). There are a number of stages in frequency filtering which are outlined below.

A forward frequency transform is applied to the image that results in two outputs: one of magnitude and one of phase. Both of these are required in order to reconstruct the image. The magnitude output for Figure 3.12a, the original input image used in this example, is illustrated in Figure 3.14a. Although this image may appear meaningless, it can be interpreted quite simply. The centre of the image represents the lowest frequency components of the original image while higher frequency components are located near the edges furthest from the centre. This image contains high-, medium-, and low-frequency components.

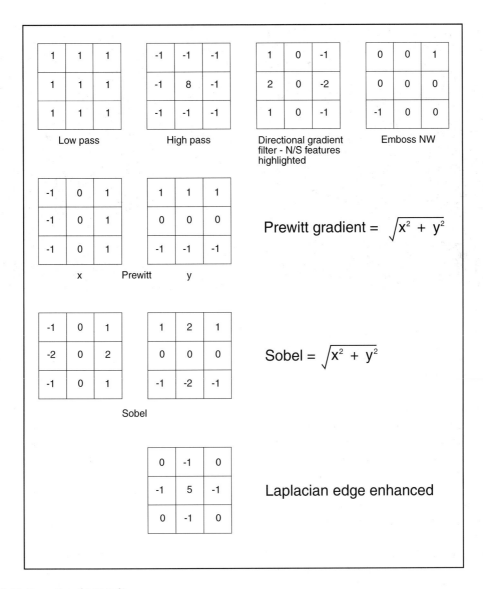

Figure 3.13 Examples of 3 × 3 filters.

To perform the equivalent of a low-pass filter, the high-frequency edges are removed from the image. This can be accomplished by producing a mask which, when applied to the magnitude image, allows only the central low-frequency components through and blocks the high-frequency ones. The methodology for producing this mask varies according to the

image processing program but one may interactively create an image with only two digital numbers, a DN of zero for the suppressed frequencies and a DN of 1 for those frequencies that are passed (Figure 3.15). If the magnitude and mask are multiplied together, a new magnitude image is created but one consisting of only low-frequency components. The application of

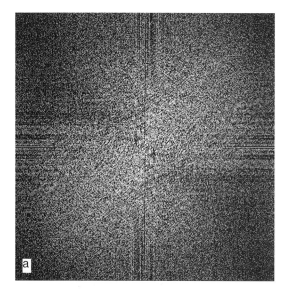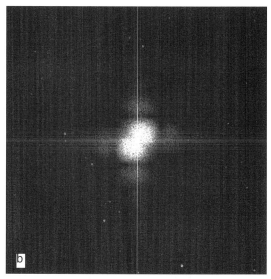

Figure 3.14 (a) Fourier image of magnitude for image shown in Figure 3.12a. (b) Fourier image of magnitude for image shown in Figure 3.12c. Note how the Fourier image for the original scene contains both the low- and high-frequency components whereas the smoothed image does not contain the high-frequency component.

an inverse-frequency transform reconstructs the image, but the new image has had the high-frequency component removed and is equivalent to a low-pass filtered image. A low-pass filter produced the image shown in Figure 3.12c. When its magnitude image (Figure 3.14b) is compared with the magnitude image for the original image, it can be seen that all the high-frequency components have been removed. It is relatively straightforward to create a high-pass filter (essentially the inverse of the low-pass one where the DN values of 0 and 1 are reversed) or a band-pass filter in which the mask consists of a ring with a value of 1 with zeros inside and outside the ring. Generally, frequency filtering is not as widely exploited as spatial filtering, though it can prove useful for reducing line-banding defects. The steps involved in frequency filtering can be time consuming and the processing time is greater. The time taken to process a 512×512 single-band image using a 133 MHz Pentium with 32 Mb RAM was 30 seconds. An advantage of frequency filtering is the large degree of control that one has over the process.

3.6 INTENSITY HUE SATURATION (IHS) TRANSFORM

Although colour images can be thought of as a combination of red, green and blue, an alternative approach, and one that the human visual system most mimics, views a colour image in terms of intensity, hue and saturation. A diagrammatic visualisation of this system is illustrated in Figure 3.16. This shows a cone, whose central axis represents intensity. Low intensities (dark) are located near the apex of the cone and higher intensity values are recorded further along the axis. The intensity values are generally represented by the digital numbers 0 to 255 in order to be compatible with most image processing systems. Hue is a measure of the colour and this can be determined by the position on the circumference. A point that plots on the circumference midway between red and green will be yellow. Saturation is a measure of the colour purity and varies from 0 along the intensity line to 1 at the circumference. It is evident from the above discussion that the intensity line is simply a measure of the grey value and is not associated with

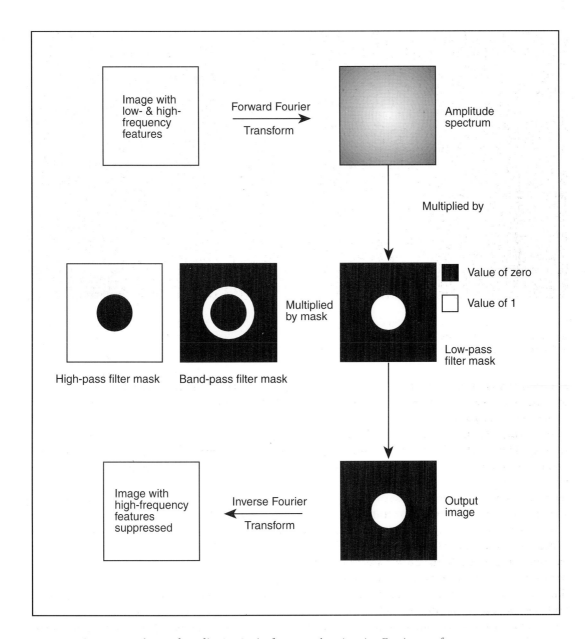

Figure 3.15 Steps required to perform filtering in the frequency domain using Fourier transforms.

any colour. A false colour composite will be relatively bland in appearance with little colour saturation if all the DNs plot close to the intensity axis. An IHS transform is an algorithm that expresses a normal red, green, blue image in terms of intensity, hue and sat-uration. Usually the saturation component is stretched, which may result in a more colourful image. It should not be assumed that an IHS image will always be superior to a conventional RGB one. The usefulness of such a transform depends on the

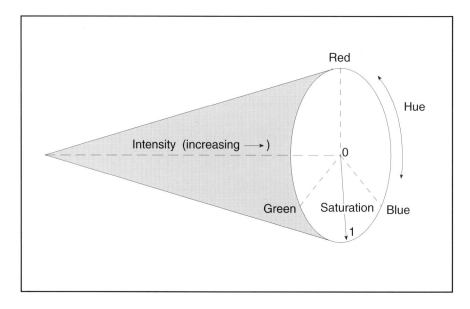

Figure 3.16 Intensity, Hue, Saturation diagram. Intensity varies along main axis of cone. Hue is a measure of colour and is plotted on the circumference while Saturation is a measure of the colour purity and increases from the centre to the perimeter.

particular features that are of interest to the analyst and the colour distribution within the scene. A false colour Landsat TM image and IHS image of part of Eritrea are displayed in Plate 3.3. The distinction between different surfaces is much more evident on the IHS image. The fine dendritic drainage patterns associated with the river system in the western half of the scene are also more prominent on the IHS image. IHS transforms can also be used to combine datasets from different systems (see section 3.8).

3.7 PRINCIPAL COMPONENTS ANALYSIS (PCA)

Although datasets may represent measurements made on one feature, they may often be correlated with other datasets. A simple example is one in which one dataset represents the age of a child while another set of data is the height of a child. One would intuitively expect that, as age increases, height would also increase. In this instance the correlation is positive because, as one parameter increases, the other one also increases. Thus, although the two datasets are sepa-

rate, they are not independent because one is able to infer one parameter given the other one. A similar situation can arise if DN values for separate bands are plotted against each other (Figure 3.17). A high correlation may exist between adjacent bands, which means the bands are not statistically independent. If a false colour composite is produced by using bands with a high correlation, the resultant image may not be colourful and pastel colours will be seen to predominate. This is because the data lie close to the achromatic line where red equals green equals blue. The human visual system is excellent at discriminating colours and, if this property is to be used to its full potential, the image being examined should be as colourful as possible. Principal components analysis is a technique which allows the production of images where the correlation between them is zero. For an n-dimensional dataset, n principal components can be produced. An important advantage of PCA is that most of the information within all the bands (represented by the variance) can be compressed into a much smaller number of bands with little loss of information. This procedure may greatly reduce the computer processing time in programs such as classi-

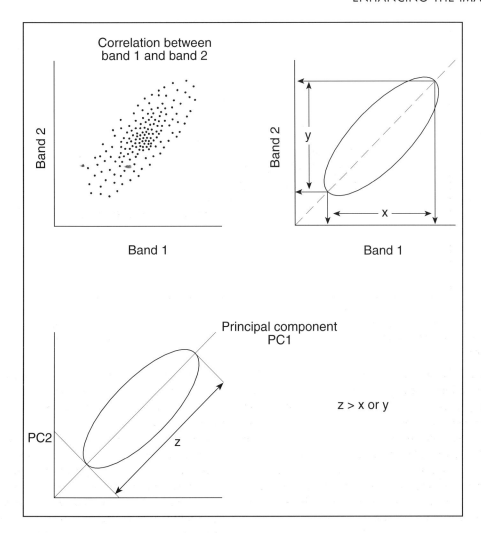

Figure 3.17 Production of principal components axes by means of a translation and rotation of the original axes. These axes are produced using eigenvalues and eigenvectors. Note how the spread of data z is greater along the first principal component (PC1) than along either of the original axes, x or y.

fication. If one considers a system with six bands and assuming all bands carry equal amounts of information (though this is not often the case), then a standard false colour composite formed from three input bands will display 50 per cent of the information in the scene. However, the first three principal components typically contain over 98 per cent of the variance. A false colour composite formed by projecting the first principal component (PC1) in red, PC2 in green and PC3 in blue thus contains most of the variance for the six input bands. Although the high principal component images (e.g. PC6) contain little variance and are usually very noisy, they should not automatically be dismissed without being examined because the information they contain may not be well represented in the lower principal components.

A principal components transform can be visualised most easily in two dimensions (Figure 3.17). The DN plot for two bands can be simply represented in a feature–space plot by an ellipse. (In three dimensions the cloud of points would form an ellipsoid.) If the bands are highly correlated, the ellipse will be very eccentric whereas, for less correlated bands, the ellipse is more symmetrical. The spread of DN values for bands 1 and 2 are x and y respectively (Figure 3.17). However, it is possible to create new axes, termed 'principal components', by means of a translation and a rotation. The long axis of the ellipse is the first principal component (PC1) and the variance (z) along this axis is greater than along either of the two input bands. The second principal component (PC2) is at right angles to PC1. The principal components transform expresses the input digital numbers in the original bands in terms of the new principal components axes. Although the visualisation of the principal components is simple, in order to create the PC axes it is necessary to calculate the length of the PC axes and their direction. These are computed by determining the eigenvalues (length) and eigenvectors (direction) from the variance–covariance matrix. The reader is directed to statistical texts such as Davis (1973) for information on how to calculate eigenvalues and eigenvectors from the variance–covariance matrix.

The principal components were calculated for the scene shown in Plate 3.4 that encompasses a number of types of land cover. The red regular areas concentrated in the top centre of the scene are forest plantations. The eastern half of the image consists mainly of cultivated fields shown in shades of pink, though there is a low-reflectance lake along the eastern edge of the image. To the west, a blue/green signature representing heather, gorse and bracken dominates the region, though dark patches show where peat has been extracted. The statistics for the six principal components formed using the six reflected TM bands (Figure 3.18) for this area are shown in Table 3.4. The first PC contains over 65 per cent of the variance while PC4, PC5 and PC6 combined have a little over 2 per cent. The six PCs are shown in Figure 3.19. PC1 is a weighted average of all the input bands; thus a pixel with the following DN – TM 1: 12; TM 2: 8; TM 3: 14; TM 4: 23; TM 5: 6; TM 7: 18 – has a DN of 33 in PC1 [(12 × 0.45) + (8 × 0.27) + (14 × 0.36) + (23 × 0.68) + (6 × 0.36) + (18 × 0.12)].

The six reflected TM bands for the scene shown in Plate 3.4 are displayed in Figure 3.18 and the corresponding six principal components are shown in Figure 3.19. Scattering affects the low TM bands (TM 1 and TM 2). The forests are sharply delineated by a dark signature that contrasts with the background vegetation on all the individual TM bands except TM

Figure 3.18 (a)

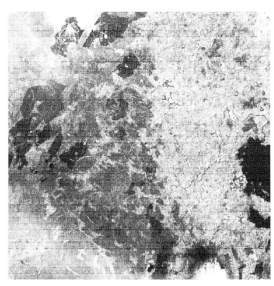

Figure 3.18 (b)

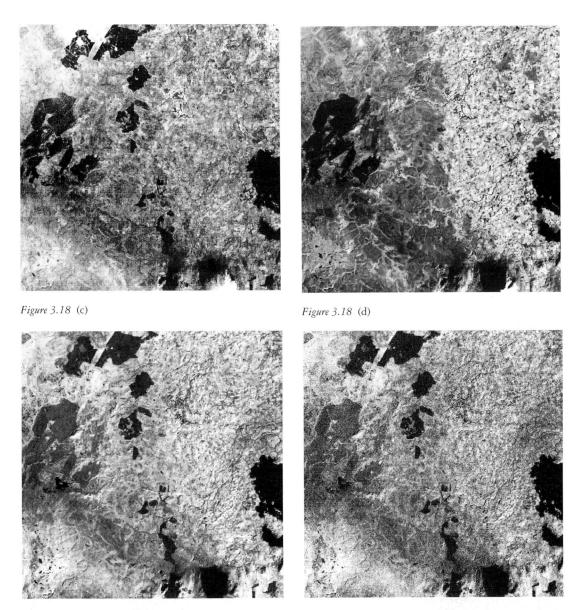

Figure 3.18 (c)

Figure 3.18 (d)

Figure 3.18 (e)

Figure 3.18 (f)

Figure 3.18 Six reflective Landsat TM images of Co. Mayo, Ireland (TM 1–5 and TM 7) used in the production of principal components shown in Figure 3.19: (a) TM 1; (b) TM 2; (c) TM 3; (d) TM 4; (e) TM 5; (f) TM 6. Data courtesy of ERAMaptec.

4, where they are associated with a high reflectivity. Differentiation of the forests (black) from the lake (mid-grey) is good on PC3. Peat cuttings have a high reflectivity in PC3 though they are most easily defined on PC1. PC1 is most like TM 4 (except for the forestry signature) because this infrared band is

Figure 3.19 (a)

Figure 3.19 (c)

Figure 3.19 (b)

Figure 3.19 (d)

the main contributor to PC1. The noise is mainly confined to principal components 4, 5 and 6 (Figures 3.19d, 3.19e and 3.19f). A false colour representation of the scene is illustrated in Plate 3.5a in which PC1, PC2 and PC3 are projected in red, green and blue respectively. Such principal component images tend to be more colourful than standard false colour composites produced from individual bands because the PC bands are not correlated. It is possible to produce false colour composites without using PC1 (Plate 3.5b). In this version, a purple signature with regular outlines is much more prominent than in the other

Figure 3.19 (e)   *Figure 3.19* (f)

Figure 3.19 Principal component images derived from Landsat TM images shown in Figure 3.18: (a) PC1; (b) PC2; (c) PC3; (d) PC4; (e) PC5; (f) PC6. Data courtesy of ERAMaptec.

false colour images. Whether or not principal component images yield significantly more information is scene dependent. The principal components image shown in Plate 3.6 can be compared with the standard false colour composite in Plate 3.2a. The PC image clearly shows the copper ore body whereas the standard false colour composite does not. The colours on a PC image cannot be directly correlated with the spectral reflectivity of different surfaces. Thus, whereas water is generally black and vegetation red on a false colour composite, water may be purple and vegetation blue on a PC colour image. It is possible to perform a decorrelation stretch on the principal components which restores the 'standard' false colours while maintaining the colour saturation.

3.8 SYNERGISTIC DISPLAY AND NON-IMAGE SPATIAL DATA

The images shown in this volume are produced from data from various remote sensing systems. However, the images have been formed by using various band combinations for a single remote sensing platform. However, different systems have different strengths and weaknesses. Landsat TM provides data in seven bands but at a coarse resolution compared to SPOT in panchromatic mode (10 m resolution) though the latter yields only a single waveband. RADARSAT, which operates in the microwave, provides very important information on roughness and texture variations but does not record differences in reflectance. Consequently, while data from a single system may provide useful information, more information can generally be obtained for any scene by combining data taken by different systems and displaying them simultaneously (synergistic display). The simplest way in which data from different systems can be viewed simultaneously is:

1   Transfer the data from the different systems into a single dataset. In order to carry out this procedure, it is important that the separate bands are co-registered with each other and that they contain the same number of lines and rows.
2   Once the data are in the correct format, a false

Table 3.4 Statistical parameters for principal components shown in Figure 3.19

Band	Mean	Standard deviation
1	36.41	5.96
2	13.12	3.14
3	13.54	4.25
4	22.20	7.41
5	17.99	5.49
7	7.39	2.09

Variance–covariance matrix

	1	2	3	4	5	7
1	35.4					
2	17.4	9.9				
3	23.0	12.6	18.1			
4	21.2	15.2	18.6	55.0		
5	4.2	4.8	7.9	26.7	30.1	
7	2.0	1.8	3.1	8.1	9.7	4.4

PC1	Eigenvalue	% Variance
1	100.4	65.69
2	37.6	24.63
3	11.6	7.61
4	1.6	1.07
5	1.04	0.69
6	0.47	0.31

Eigenvectors of covariance matrix

	PC1	PC2	PC3	PC4	PC5	PC6
Band 1	0.45	0.62	0.18	−0.6	0.02	0.14
Band 2	0.27	0.23	0.02	0.23	0.01	−0.9
Band 3	0.36	0.30	0.25	0.75	−0.10	0.38
Band 4	0.68	−0.33	−0.65	−0.01	0.05	0.09
Band 5	0.36	−0.57	0.63	−0.16	−0.33	−0.07
Band 7	0.12	−0.17	0.28	0.04	0.94	0.02

colour composite can be formed by projecting three bands in red, green and blue.

Plate 3.7 illustrates an image in which TM band 2 is projected in red, the ratio of TM 3 to TM 4 is pro-jected in green and a radar image is projected in blue. The inclusion of radar data, which are particularly useful for providing textural information, comple-ments the spectral information that is provided by Landsat. Which specific bands are combined depends to a great extent on what the operator wishes to high-light and on the characteristics of a scene. A botanist wishing to study vegetation may include near-infrared datasets whereas a geologist studying imagery in which desert varnish masks the natural reflectance of the rocks may opt for radar and thermal bands. It is also possible to produce synergistic dis-plays using data that were obtained in non-image for-mat as will be discussed later in this section and in Chapter 5.

Simply projecting three bands in red, green and blue does not increase the spatial resolution of the image. However, it is possible to combine panchro-matic SPOT data with multispectral TM data in order to yield a false colour image with an apparent 10 m spatial resolution. The steps involved are:

1 Produce a false colour composite TM image with a spatial resolution of 30 m.
2 Perform an Intensity Hue Saturation (IHS) trans-form on the TM image.
3 Substitute the 10 m resolution panchromatic SPOT image for the intensity component.
4 Transform the IHS image back onto the red, green, blue axes. This procedure results in a false colour image in which the hue and saturation are controlled by the original input TM image, but the spatial resolution is that of the panchromatic SPOT image.

Some image-processing software (PCI and ERMap-per, for example) has specially designed programs which perform the fusion of such data automatically. Plate 3.8 shows an airport in California on a TM image and also on a SPOT/TM merged image. The runways are much better defined on the merged image.

Some of the natural properties of any area can be measured by remote sensing methods and then con-verted into an image. SPOT, for example, measures the reflectance characteristics in three wavebands (four for SPOT 4) which can then be displayed as three grey-tone images. Similarly, band 6 on Landsat TM allows the production of a grey-tone image which shows variations in the thermal characteristics of a scene. However, there are other properties for a region, for which none of the polar-orbiting satellites such as Landsat, SPOT, IRS and JERS obtains data.

One of the most fundamental properties of a region is its topographic expression. Elevation data are often collected at widely scattered points and displayed as a contour map. In order to produce such a map, the irregularly scattered points are interpolated onto a regular array from which the contour levels are drawn and then the array is discarded (Figure 3.20). However, as the array is identical in format to a digital image, it is possible to produce a digital elevation model (DEM). Once the data are in a digital format, they can be manipulated by the various enhancement procedures discussed earlier. Figure 3.21a illustrates an elevation image for part of California in which the lower the location, the darker the image. Although

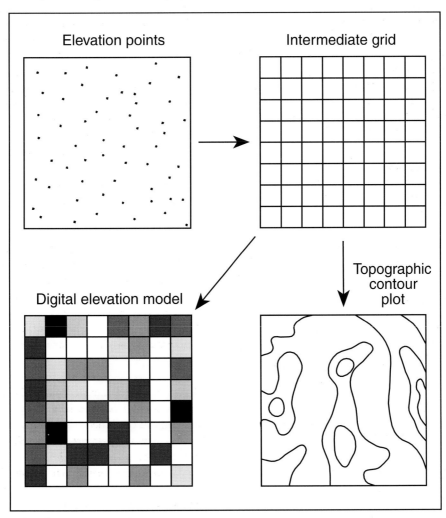

Figure 3.20 Formation of an image using irregularly spaced point data not collected by a remote sensing sensor.

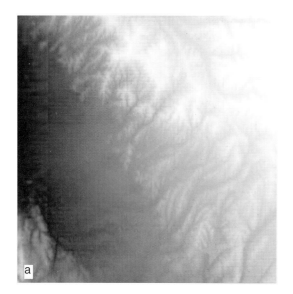

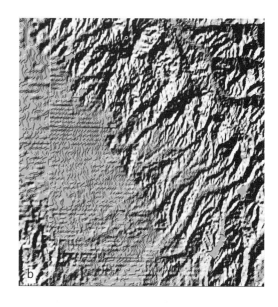

Figure 3.21 (a) Elevation data shown in image format. Pseudo-illuminated versions of (a) with illumination from (b) the southeast and (c) the southwest. Digital imagery provided by Space Imaging EOSAT Inc.

the image provides information on elevation differences, it is relatively bland. The human visual system evolved to process three-dimensional information and it is possible to convert the flat image into a three-dimensional representation by providing pseudo-illumination from a specific direction. Generally, the best presentation uses low illumination angles with the illumination coming from the northeast (45

degrees), southeast (135 degrees), southwest (225 degrees) or northwest (315 degrees). Two versions of Figure 3.21 are shown, both of which have illumination angles of 20 degrees, with illumination directions from the southeast (Figure 3.21b) and the southwest (Figure 3.21c). The elevation differences for the scene are much more obvious on the pseudo-illuminated images and illustrate that a mountainous region exists to the north of the scene and is paralleled by mainly flat low-lying terrain in the southern part of the area.

It is possible to include such non-image data within synergistic displays. Plate 3.9 illustrates a false colour image produced by 'draping' TM and radar data over a digital elevation image in order to create a perspective view. The image shows that the field systems are confined to the low-lying regions. Many image processing systems also allow the view to change sequentially as if the operator is flying through the terrain.

3.9 CHANGE DETECTION ANALYSIS

A significant advantage of remote sensing systems over conventional systems for environmental monitoring is the repetitive coverage provided by spaceborne platforms. The repeat cycle for such platforms varies greatly, but for most polar-orbiting satellites with spatial resolutions of less than 100 m, it is of the order of 20 days. Such satellites have been operational for many decades, and thus provide a record of change and – often of greater importance – rates of change of the environment. These changes may simply reflect natural cycles but are often caused by human activity. Change detection has a number of applications within the environmental sphere. Urbanisation is continually encroaching on the green belts surrounding many cities and the growth and direction of urban development may be monitored. Land-use change may entail documenting changes in the types of crops being grown in a particular region or alternatively mapping changing field patterns as hedge boundaries are removed in order to create larger fields for more efficient use of machinery. The extent of deforestation in tropical rainforests can also be mapped. The possibility of global warming is an environmental issue that is currently under investigation by scientists in a range of disciplines. One signal of increasing temperatures that can be readily monitored from spaceborne remote sensing systems is the retreat of valley glaciers. Landsat images from 1973 and 1986 show that in that time period the Muir glacier in Alaska retreated by 7 km.

Two simple arithmetic procedures, subtraction and addition, are used to produce a change detection (or difference) image. Initially two images for the same scene are obtained and co-registered to each other. It is important that the registration be as accurate as possible and with an accuracy of less than one pixel. Change detection, like ratioing, is performed on a pixel-by-pixel basis; a high accuracy is thus required in order that the pixels match. An interim stage is the subtraction of the digital numbers for one image from the digital numbers for the other image (Figure 3.22). However, this procedure may yield negative digital numbers that cannot be displayed on conven-

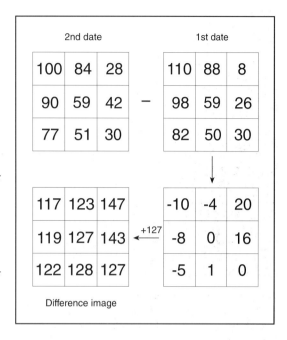

Figure 3.22 Construction of a change detection (difference) image.

tional digital image processing systems. A constant, usually 127, can be added in order to achieve a positive range of digital numbers. In theory, the digital numbers for the interim image may range from -255 to $+255$, though in general the range tends to be much smaller. However, there are situations where large numbers can be obtained. A river course, which will have a very low DN when water is present, may be associated with a very high DN when it is dried up if it is delineated by high reflective sand such as occurs in north Africa. As a separate difference image can be produced for each band, TM data can yield seven difference images. The interpretation of a grey-tone difference image is relatively simple. Those parts of the scene which have not changed between the two dates are shown as a mid-grey colour, while the darkest and brightest parts of the scene show the maximum changes (Table 3.5). It is also possible to create false colour change detection images by projecting three difference bands in red, green and blue, though the interpretation of such images requires careful consideration.

Table 3.5 Colour examples on an 8-bit false colour change image

Red		Green		Blue		
DN (1)	Change (1)	DN (2)	Change (2)	DN (3)	Change (3)	Colour
127	none	127	none	127	none	mid-grey
255	max.	255	max.	255	max.	white
0	max.	0	max.	0	max.	black
0	max.	255	max.	0	max.	green
127	none	0	max.	127	none	dark purple
0	max.	127	none	127	none	turquoise

A black or white tone on a false colour composite represents areas of maximum change whereas a mid-grey tone shows no change (a DN of 127 in each of the three bands). A green, in the example shown in Table 3.5, is due to a maximum change in all bands but in this instance the maximum change for band 2 is 255 while for bands 1 and 3 it is 0. A dark purple can be formed by there being no change in the bands projected in red and blue (DN of 127) and a maximum change in green (DN of 0). A dark grey signature represents a certain degree of change between the images.

Although the concept of a change-detection image is relatively simple, there are a number of important factors that must be borne in mind. The premise underlying change detection is that any changes in digital number between the dates being considered are caused solely by changes in the spectral reflectance of the surface, presumably because of some alteration that has occurred. However, it may not always be possible to exclude all other sources of DN variation totally. As an example, consider a scenario in which a researcher wishes to examine the changes between 1990 and 1998. It is very important that the images for 1990 and 1998 be obtained if possible on the same calendar date. Images obtained at different times of the year may appear dissimilar even if no change has occurred because the illumination angle and Sun–Earth distance varies throughout the year (see section 2.8). However, it may not be possible to obtain images taken on the same calendar date in different years for a number of reasons. First, the repeat cycle for the satellite may be such that on the same calendar date the satellite may not be imaging the same area. Second, even if the anniversary date and the satellite's overpass do coincide, there may be extensive cloud cover for one of the dates. Obtaining images that are cloud free for the same area taken on the same calendar date in two given years may prove very difficult. One is often forced to approach a change-detection exercise from the opposite direction and ascertain what cloud-free images are available for a specific area extending back a number of years and then decide whether any of the available data are suitable for the investigation being undertaken. Thus, multitemporal December images may be suitable for an urban planner to ascertain the degree of urban development that has occurred but to a botanist who wished to ascertain changes in vegetational patterns a December image may prove totally unacceptable because of the dieback of vegetation in the winter. In such a situation the botanist may have to accept images taken in different months of the year. Atmospheric conditions may vary between the two dates and the presence or absence of clouds for the same area is the most obvious manifestation of this. As discussed in Chapter 2, the DN recorded by a sensor has an atmospheric component; the DNs associated with a single pixel may therefore differ, not because of a change in the reflectance of the surface but because of different atmospheric components. Variations in the atmosphere are more likely to affect the shorter wavelength bands. Table 3.6 lists some of the statistical characteristics of the datasets used to produce the images shown in Plates 3.10a and 3.10b, which show two Landsat MSS images, one taken on 6 May 1989

Table 3.6 Statistical parameters for a site imaged on 2 May 1990 and 6 May 1989 by Landsat MSS sensors

	Minimum	Maximum	Mean	Standard deviation
2 May 1990				
Band 1	37	255	55	7.5
Band 2	21	255	41	11
Band 4	19	255	151	44
6 May 1989				
Band 1	43	171	64	7
Band 2	27	150	45	10
Band 4	21	255	153	40

and the other on 2 May 1990. The longer infrared waveband (band 4) is most similar between the two dates whereas band 1 appears most different.

Although it may not be possible to obtain two datasets taken on the same calendar date in different years by the same remote sensing system, it may be possible to compare data that do meet this criteria by using data obtained by different systems. The large number of remote sensing platforms currently in operation increases the likelihood that suitable data can be obtained, though again a number of problems may arise in any comparison. Landsat and SPOT together have provided the remote sensing community with a large amount of data. However, it is not a simple matter to compare directly a suitable SPOT and Landsat image. A typical 60×60 km scene contains about 9 million pixels on SPOT (multispectral) and 4 million on Landsat TM. Consequently, one cannot simply perform the simple subtraction illustrated in Figure 3.22. To compare SPOT and TM would involve resampling the SPOT data to a 30×30 m pixel size in order to make them compatible. Since the reflectance of a surface varies with wavelength, it is important, when subtracting one image from another, to use the same band. Thus band 1 on one date is subtracted from band 1 on another date. This is not a problem if a single system is used but other difficulties arise if data from different systems are used. In general the wavebands used for different

platforms vary, so that one is not comparing reflectances obtained for the same part of the electro-magnetic spectrum. If one wished to compare a TM 4 (0.76–0.90 μm) image with SPOT, the comparable waveband is band 3 (0.79–0.89 μm). Using JERS, IRS and ADEOS, the comparable wavebands are 0.76–0.86 μm, 0.70–0.90 μm and 0.76–0.89 μm respectively. Another complication is that, while Landsat images the terrain directly beneath the satellite, SPOT can obtain oblique views of an area; one should therefore, if possible, use only images obtained at nadir if a comparison of TM and SPOT is to be attempted.

The anniversary dates for the images in Plate 3.10 are very close, and therefore variations in the Earth–Sun and illumination angle and direction are low. Azimuth direction angles and illumination angles both vary by only two degrees between the images. Although the images shown in Plate 3.10 are of the same region, they are in fact subsets of two full MSS scenes, which have the same row number (23) but adjacent path numbers (206 and 207) (Figure 3.23). The large overlap between the images in adjacent paths allows the changes within this area between 1989 and 1990 to be evaluated. The false colour images taken on both dates appear very similar and are dominated by two main signatures. In this region in May, the fields contain crops or pasture and yield bright red colours due to the high reflectance in the infrared. Fallow fields are blue and are mainly concentrated in the east central parts of the images. The Slieve Bloom mountains (west central) are generally not farmed as intensively as the lowlands: they are often scrub and heather covered and are associated with a dark green colour. There is a similar signature in the northeast from peat working, where the covering vegetation has been removed. There are various approaches to displaying difference images. Jensen (1996), for example, suggests displaying one date in red and the other in green. Varying shades of yellow will then represent areas that have not changed between the two dates. Grey-tone difference images for the three separate bands that comprise the false colour composites are shown in Figure 3.24. A larger tonal variation between the two years is shown for the

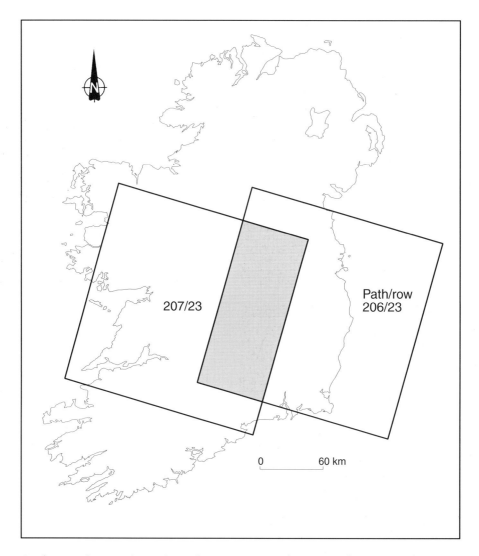

Figure 3.23 Overlap area, showing where a change detection image can be generated by using data obtained on different paths.

band 1 difference image (1990–1989), possibly because of the noticeable haze component on the 1989 image. Although grey-tone images may be employed, it is also possible to density slice the images. Plate 3.11 illustrates a density-sliced version of Figure 3.24c using the colour codes given in Table 3.7.

Those areas that have remained reasonably con-

stant between the two dates are displayed as black. On the band 4 density-sliced difference image, those areas which are unchanged are the Slieve Bloom mountains and the regions of peat cutting in the northeast. The remainder of the region is displayed in green and blue. These areas correspond to the cultivated fields of the lowlands; most of the changes are thus caused by variable spectral responses for the

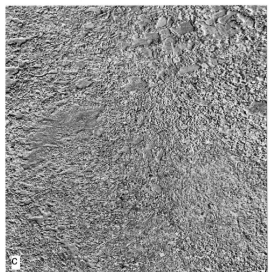

Figure 3.24 Grey-tone change detection images for (a) Landsat MSS 1; (b) Landsat MSS 2; (c) Landsat MSS 4.

same calendar date could be showing vegetation at different stages in its phenological cycle. These differences would be best displayed on an image which acquired data in the infrared, such as MSS band 4. An examination of the monthly weather bulletins for this region shows that in April 1989 (the month preceding the acquisition of the 1989 image) it was cold and frosty (temperatures below average) with higher than normal rainfall being recorded at most measuring stations.

Table 3.7 Colour code for density-sliced Landsat MSS 4 image shown in Plate 3.11

Range	Colour
More than 3 standard deviations lower	dark blue
2–3 standard deviations lower than mean	pale blue
1–2 standard deviations lower than mean	green
±1 standard deviation	black
1–2 standard deviations higher than mean	yellow
2–3 standard deviations higher than mean	orange
More than 3 standard deviations	purple

crops. There are a number of possible explanations for this pattern. One is that different crops are being grown in the fields on a yearly rotation cycle. However, given the widespread nature of the different signatures, it is unlikely that this could be the full explanation. It is more likely that climatic conditions were different in the two years. For example, a cold spring could either delay the planting of crops or retard their growth, so that images obtained on the

3.10 CLASSIFICATION

An image is different from a map in one fundamental respect. Maps generally represent one surface category by a specific colour or symbol. Thus, for example, a map may show water as a uniform blue colour. However, on remotely sensed images, a range of digital numbers rather than a single value represents a single surface class such as water. Many maps have large areas that are blank (or white), i.e. no land use or land cover has been assigned to them. Images, in contrast, provide a continuous record of land cover in these blank areas. However, units of land are often complex, with several cover types contributing to the observed signature. Classification is therefore often used to smooth out small, insignificant variations and simplify an image into a thematic map of land cover.

Classification is the process by which pixels which have similar spectral characteristics and which are consequently assumed to belong to the same class are identified and assigned a unique colour. Once an image is classified, it is possible to interrogate the dataset and ascertain the area of different classes. Thus one may classify an agricultural region in order to determine the areal extent of different crops. Figure 3.25a shows how the reflectance varies with wavelength for three hypothetical surfaces that are being imaged by a system that obtains data in three bands. The three surfaces cannot all be differentiated from one another on any single band. Thus surfaces B and C appear similar on band 1 (i.e. they have the same DN), surfaces A and B are similar on band 2 and A and C are similar on band 3. However, if the three pixels are represented on a feature–space plot produced from the three bands, they are seen to be clearly distinguishable from each other (Figure 3.25b). In general, a more accurate classification result is obtained the greater the number of bands that are used. Increasing the number of bands to be used in classification also greatly increases the computing time. In this situation one may opt to classify the principal component images. In the example provided earlier in this chapter, the first three principal components held 97.93 per cent of the available variance. Another approach that may be adopted if computing time is a factor is to use only the bands that show the lowest correlation with each other. The example shown in Figure 3.25b is very simplified because the DN for a given surface will in reality encompass a range of digital numbers. Thus, instead of surfaces A, B and C being represented by points in the feature–space plot, they will occupy discrete volumes which may overlap (Figure 3.25c). The process by which a pixel is assigned to a particular class is based on the statistical characteristics of the image and the pixel. There are two main forms of classification: supervised classification and unsupervised classification.

Supervised Classification

Supervised classification involves a considerable amount of input from the image analyst and a knowledge of the types of surface that are found in the study area. This information can be obtained from maps or from actual fieldwork where different surface classes are identified and their geographical positions noted. The classification process involves a number of steps (Figure 3.26).

Initially the operator projects the image to be classified onto the display monitor and outlines sample or training areas for each surface class. These are used to provide the classification program with typical examples of each kind of land cover to be used in the classification. The computer generates statistical parameters from the training areas and compares the digital numbers of every pixel in the image with these statistical parameters. If the DNs for a pixel fall within a known training area, then the pixel is assumed to belong to the same surface class as the training area. (The algorithms used in the comparison are discussed later.) After the classification process has taken place, different colours represent different surface classes. Although the general process is relatively simple, it is worth considering in detail some of the problems that can be encountered. A sufficient number of pixels for each surface class must be delineated in order to ensure that a representative sample is obtained for each class. The training areas for any one class should not be concentrated in one part of the

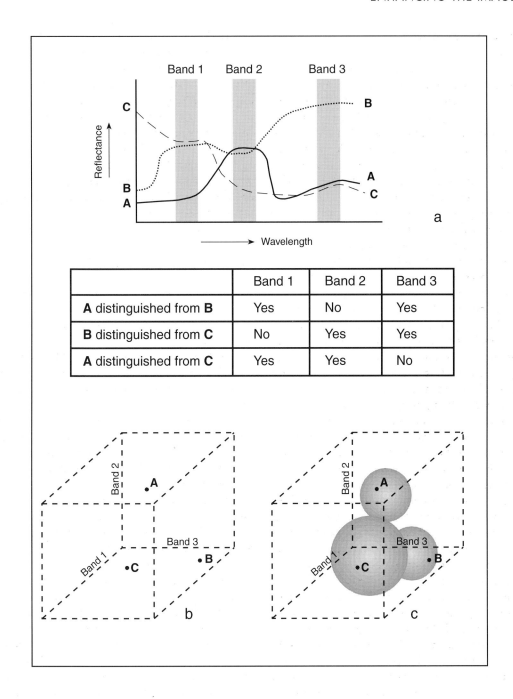

	Band 1	Band 2	Band 3
A distinguished from **B**	Yes	No	Yes
B distinguished from **C**	No	Yes	Yes
A distinguished from **C**	Yes	Yes	No

Figure 3.25 (a) Differentiation of surface classes A, B and C in bands 1, 2 and 3. No single band allows all the surfaces to be distinguished. (b) Viewed in a three-dimensional framework the surfaces can be separated but (c) in reality the surfaces will occupy discrete volumes in the feature-space plot which may intersect.

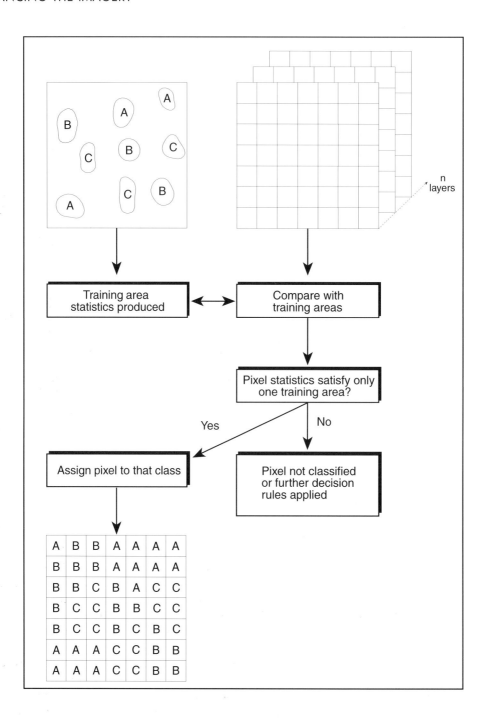

Figure 3.26 Generalised sequence of steps in a classification process.

image but should encompass the entire scene. The histograms for training areas should be unimodal and conform to a normal distribution (Campbell 1996). The training areas should be as separate and uniquely representative as possible, otherwise a substantial overlap between classes may occur and pixels will be misclassified. It may not even be possible to ensure that classes are discrete because they may have similar reflectance characteristics in the bands that are being classified. In such a situation it may be preferable to merge the training sites and consider them as a single class. The false colour composite shown in Plate 3.12a was classified using seven known land classes and the resultant classified image displayed in Plate 3.13a, where the colour key shown in Table 3.8 applies.

The locations of the training areas (with the same colours as in Table 3.8 and with the image blacked out for ease of viewing) are displayed in Plate 3.12b and the statistical parameters for the training sites given in Table 3.9. (Only three bands were used in this example though in practice more bands are usually employed).

It may be preferable to isolate an individual class and this is simply achieved by assigning a value of zero to all other classes (Figure 3.27). Table 3.9 is not particularly informative if a decision is required about whether the training areas are sufficiently different for a meaningful classification. There are a number of methods which allow the operator to decide whether the training areas are sufficiently different. Plate 3.13b shows feature–space plots for three two-band combinations which allow compar-

Table 3.8 Colour code for classified image shown in Plate 3.13a

Colour	Land class	% of image
Pink	water	0.07
Pale blue	oil seed rape	0.12
Green	late crops	30.21
White	bare soil	12.4
Yellow	urban/industry	7.2
Blue	forest/early crops	8.54
Red	pasture	18.63
Black	unclassified	22.83

Table 3.9 Statistical parameters of training areas

	Mean	Standard deviation
Training area: water		
TM 3	18.6	1.5
TM 4	13.5	4.7
TM 5	8.9	4.1
Training area: oil seed rape		
TM 3	62	7.4
TM 4	150	2.6
TM 4	67	2.7
Training area: late crops		
TM 3	24.7	1.3
TM 4	123.2	8.5
TM 5	86.3	3.9
Training area: bare soil		
TM 3	53.6	11.7
TM 4	82.3	14.2
TM 5	119.5	13.1
Training area: urban/industry		
TM 3	35.4	4.9
TM 4	58.8	11.3
TM 5	65.5	7.4
Training area: forest/early crops		
TM 3	20.4	1.5
TM 4	92.6	9.3
TM 5	47.9	7.1
Training area: pasture		
TM 3	20.7	1.1
TM 4	142	7.6
TM 5	68	6.2

isons of the amount of overlap between different training sets. The white 'cloud' within the soils ellipse shows the DN values which have been interactively outlined to provide the training area. The largest concentration is around the mean in the centre of the ellipse. The operator controls the size of the ellipse. In the example shown, the axes of the ellipses are three standard deviations long. Decreasing this value reduces the degree of overlap. A number of statistical measures, which are often provided with the image processing software, can be computed in order to yield a separability matrix for the training areas. One is known as the Bhattacharrya (or

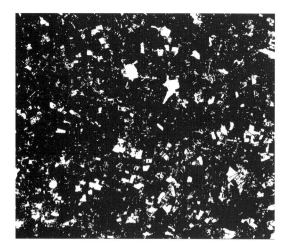

Figure 3.27 Isolation of a single surface class (soils) by assigning a DN of zero to all other classes.

Jeffries–Matusita) distance. The details of this measure are beyond the scope of this book but one may consult Richards (1993) and the references therein. This measure yields values ranging from 0 to 2 (Table 3.10). A value of 0 between two training areas means there is no separability and the training areas are identical. Values between 1 and 1.9 indicate poor separability but values greater than 1.9 show good separability. The training sets chosen for this exercise all show good separability and thus each may be considered to represent a distinct class.

The lowest separability (1.93) is between pasture and late crops. Although the above parameters indicate that the training areas appear to have been well chosen, an examination of Table 3.8 shows that approximately 23 per cent of the image remains unclassified, i.e. the DN distribution for these unclassified pixels did not fall within the parameters set by the training areas. This could be because there is another distinct class present in the scene which has not been considered or the unclassified pixels belong to the known classes but the training areas did not encompass their DN distribution or a combination of the two. An examination of the position of these pixels shows that they represent the boundaries between known classes. Thus, in the example given, it is apparent that the chosen training areas did not wholly encompass the full range of DNs for any given class. Figure 3.28 shows some of the difficulties that may arise. At the boundary between two classes, a pixel's DN is formed of a combination of two surface classes. However, this pixel (sometimes referred to as a mixel) in conventional classifications can be assigned to only one class and the resultant classification might yield Figure 3.28b. Some areas will consequently be misclassified. Further complications may arise because (a) the boundary between the two classes may be delineated by a hedge or other physical barrier with its own distinct signature and (b) the reflectance values change gradually rather than abruptly as one moves from one surface class to another. Consequently, if the training areas are as shown in Figure 3.28a, then a zone of unclassified pixels may be produced between them (Figure 3.28c). Including training areas from near the edges may rectify this but this will in turn probably decrease the separability of the classes.

Table 3.10 Separability matrix for training areas

	Water	Oil seed rape	Late crops	Bare soils	Urban	Forest/early crops
Oil seed rape	2.00					
Late crops	2.00	2.00				
Bare soil	2.00	2.00	1.95			
Urban	1.99	2.00	1.99	1.97		
Forest/early crops	2.00	2.00	1.99	1.99	1.94	
Pasture	2.00	1.99	1.93	1.99	1.99	1.99

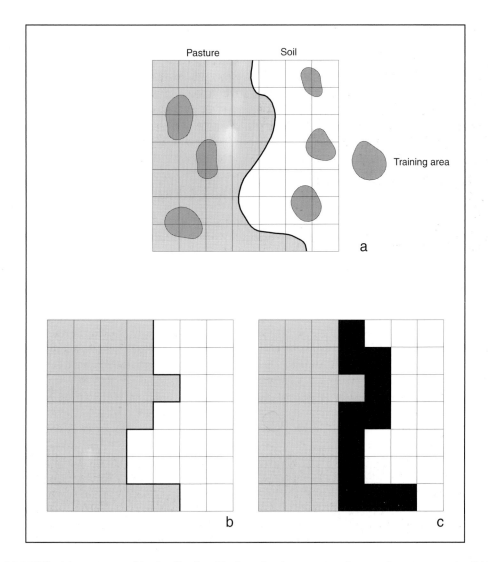

Figure 3.28 Difficulties encountered in classification. The boundary between two classes such as pasture and soil (a) may be sinuous but the resultant classification boundary may be (b) blocky or (c) separated by a narrow zone which does not fit into either class.

Unsupervised Classification

Unsupervised classification is a technique that groups the pixels into clusters based upon the distribution of the digital numbers in the image. An unsupervised classification program, such as ISODATA clustering, may require the operator to specify a number of parameters such as:

- a maximum number of classes,
- a maximum number of iterations,
- a threshold value.

An unsupervised classification operates in an iterative fashion. Initially the computer program assigns arbitrary means to the classes and allocates each pixel in the image to the class mean to which it is closest. New class means are then calculated and each pixel is then again compared to the new class means. This procedure can be repeated as many times as the number of iterations input into the program. However, pixels move between clusters following each iteration. Once the user-defined threshold is reached, the program terminates even if the maximum number of iterations has not been reached. A threshold of 0.98 means that the program terminates when less than 2 per cent of the pixels move between adjacent iterations. The classes produced from unsupervised classification are spectral classes and may not correlate exactly with 'information classes' as determined by supervised classification. Plate 3.14 shows the results of an unsupervised classification using approximately the same colours as those used in Plate 3.13a. This program has classified the boundaries between the surface classes and has generally assimilated them into the class shown green. Figure 3.29a shows the water signature (white) produced by supervised classification to allow a comparison with a class produced by the unsupervised program (Figure 3.29b). The lat-

ter identified known rivers in the southeast, which the supervised classification failed to pick up. However, ground investigations have shown that this spectral class also includes known urban and forest/early crops areas. If a similar exercise is performed for known urban regions (Figure 3.30a), then the nearest equivalent in an unsupervised format (Figure 3.30b) locates most of them but also includes some non-urban areas.

Classification Algorithms

In supervised classification, each pixel is compared with each training area and is assigned to the class to which it is most similar. In unsupervised classification, a pixel is also assigned to the cluster to which it is closest. However, there are a number of statistical algorithms which can be used to measure the extent of the similarity between a pixel and a class and different classifiers may assign the same unknown pixel to a different class. The simplest to understand is known as the minimum distance (or nearest neighbour) classifier. Figure 3.31a shows the position of three surface classes (A, B and C) in a two-dimensional feature space plot. The cloud of points increases towards the centre of each class, which can

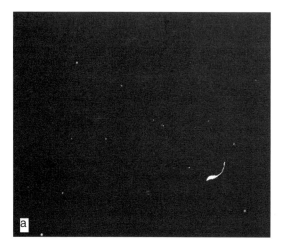
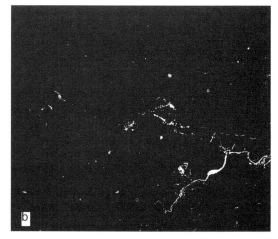

Figure 3.29 (a) Information class 'water' shown for study area, which can be compared with (b) spectral class obtained by unsupervised classification. The latter has identified more water bodies than the image obtained by supervised classification, but ground investigations show that some pixels in this spectral class are urban and forest regions.

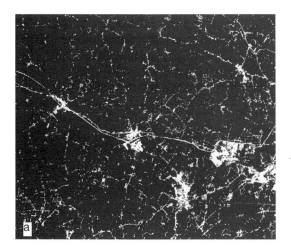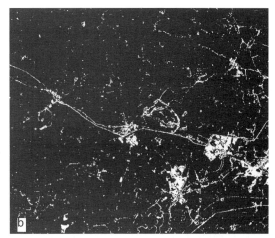

Figure 3.30 (a) 'Urban' information class which is similar to the (b) spectral class shown, though the latter also includes non-urban regions.

be equated as the mean of the class. Unknown pixels, such as 1, 2, 3 and 4, are also plotted and assigned to the class to which they are closest (that is the shortest distance from the mean of the class to the position of the point in a feature space plot). Unknown pixels 1 and 2 are assigned to class A by the minimum distance classifier, while pixels 3 and 4 are assigned to class C. A parallelepiped (or box) classifier constructs boxes around the class. In the example shown in Figure 3.31b, the box encompasses the maximum and minimum values but the size can be varied in standard deviation units either side of the mean. A pixel is considered as belonging to a particular class if it falls within the box. A pixel that does not fall within any box is not assigned to a class. Pixels may fall within two boxes if the boxes overlap. In this situation they may be left unclassified or classified according to specified decision rules. When the box classifier is used, pixel 1 falls within two classes and may be left unclassified, pixel 2 is assigned to class B while pixels 3 and 4 are unassigned. A maximum likelihood classifier (which was used to produce Plate 3.13a) assumes a normal distribution for the training areas. Probability contours are created around each training area and a pixel assigned to a class depending upon the value of the probability contours that encompass it (Figure 3.31c). The maximum likeli-

hood classifier is generally considered to be the most powerful but is also considered the most computer intensive. According to this algorithm, pixel 1 belongs to class A, pixel 2 to class B and pixel 4 to class C. Pixel 3 has a higher probability of belonging to class B than class C. (Note that the dispersion for the pixels in class B is greater than in class C; an unknown pixel equidistant from both therefore has a greater affinity with class B). It is possible to introduce a 'weighting factor' into the classification process if there is a priori information that one class has a greater probability of being present than another class. For example, two surface classes A and B may have similar DN distributions, but it may be known that B is very rare within the study area. Thus a higher probability weighting factor may be assigned to A. The term 'Bayesian classifier' is used for a maximum likelihood classifier in which the probability weightings are different for the classes.

Advantages and Disadvantages of Supervised and Unsupervised Classification

An advantage of unsupervised classification is that no prior knowledge of the scene is required but the classes found by this technique cannot be attributed

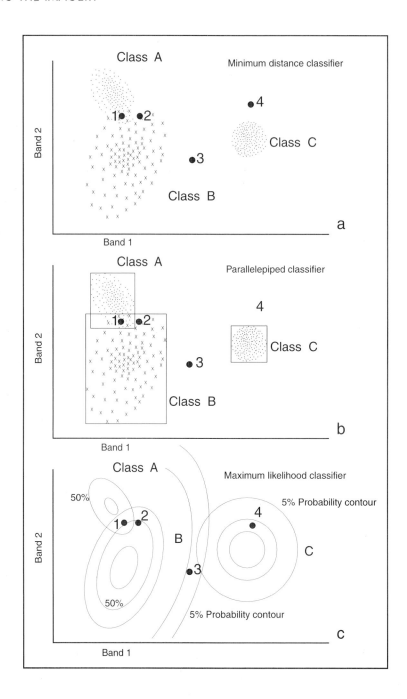

Figure 3.31 Classification procedures: (a) minimum distance, (b) parallelepiped and (c) maximum likelihood.

to a particulate type of surface without some independent information from a ground investigation or from published sources. The 'set-up' time is much shorter for unsupervised classification, which generally requires a minimum input from the operator, whereas training sets have to be identified, delineated and assessed before supervised classification can proceed. However, although this may be more time consuming, the resultant classification may be more useful because particular classes can be attributed to specific types of surface. Supervised classification works on the premise that the training areas are representative of a particular surface class, so that the onus is on the operator to ensure that this condition is realised. Poorly chosen training areas will yield a poor classification.

Refinements on the Classification Procedure

The conventional classification procedures discussed in earlier sections, which use the spectral characteristic of the pixel as the sole parameter in deciding which class a pixel belongs to, can be improved by a variety of techniques. Classification is akin to ratioing in that each pixel is taken in isolation without any reference to the pixels around it. However, the classification may be improved by considering a pixel in the context of its neighbouring pixels (contextual classifier). This approach is best described by a simple example. A pixel would only be assigned to a 'beach' class if it had the spectral signature of the beach class training area *and* it was within a specified distance of water. Other ancillary data may also be incorporated in order to improve the classification. Certain types of vegetation may be altitude sensitive, so that incorporating a digital elevation model may prove advantageous. If a file containing field boundary vectors is available, the operator may insist that only one surface class (presumably a specific crop) may be allocated to a single field. Classified images may be 'noisy', with isolated pixels of one class surrounded by pixels of another class. A mode filter can be applied to the thematic image that replaces the isolated pixel by the most frequently occurring class within the filter window.

As mentioned earlier, conventional classification assigns a pixel to a single class. Jensen (1996) refers to this as 'hard classification'. If a pixel is included within one class, it is by definition excluded from all other classes. However, in the real world pixels may include a number of classes and the spectral signatures for these mixed pixels are due to a combination of the classes located within the pixel. The simplest situation is when only two classes are involved (see Figure 3.28a). There are two 'end-members' in Figure 3.28a, a pixel that is wholly pasture or wholly soil. However, some pixels consist of different proportions of soil and pasture. More complicated situations will also occur, especially for remote sensing systems with low spatial resolutions. A pixel that is 100 m square is liable to encompass more classes than a pixel that is 5 m square. Fuzzy classification is a technique in which each pixel is assigned a number for each class, ranging from 0 to 1, which indicates the proportions of the different classes which have contributed to the observed spectral signature. Table 3.11 shows a sample result. Pixels A and C would have been assigned to water and bare soil respectively by conventional classification and pixels B and D would possibly not have been classified. However, the fuzzy classification shows that pixel B contains approximately equal amounts of water, forest and crops while pixel D is formed mainly of bare soil and pasture. Some image processing software incorporates programs which use the fuzzy logic approach to classification. Initially end-member training areas are chosen from the image display which are known to be spectrally pure

Table 3.11 Typical classification result using fuzzy logic

Class	Pixel A	Pixel B	Pixel C	Pixel D
Water	0.9	0.3	0.2	0.0
Oil seed rape	0.0	0.0	0.0	0.1
Pasture	0.0	0.0	0.0	0.4
Bare soil	0.1	0.0	0.8	0.4
Urban	0.0	0.0	0.0	0.1
Forest/early crops	0.0	0.4	0.0	0.0
Late crops	0.0	0.3	0.0	0.0

(signature is due to the presence of only one class). Each pixel is then classified, assuming the spectral signature is due to a linear combination of all end-members. If there are n input end-members then n output images are produced. Thus a soils output image (assuming that soils was an end-member) would be dark for pixels in which soils did not contribute to the signature and very bright for pixels where it made a significant contribution. A limitation to this program is that the number of end-members that can be employed must be less than or equal to the number of input bands. Thus, if the six reflective Landsat TM bands are used, a maximum of six classes can be employed.

Neural network classification algorithms are increasingly being used in remote sensing. Unlike the maximum likelihood classifier, they do not rely on the assumption that data are normally distributed. A surface class may be represented by a number of clusters in a feature space plot rather than a single cluster (Atkinson and Tatnall 1997). A special issue of the *International Journal of Remote Sensing* (volume 18, number 4, 1997) considers a range of neural network applications in remote sensing and the reader is directed to this publication for further discussion.

3.11 IMAGE PROCESSING HARDWARE AND SOFTWARE

Introduction

Although different digital image processing procedures have been discussed in this chapter and Chapter 2, so far there has been no discussion of the types of hardware and software necessary to accomplish them. Image processing hardware and software have been in existence since the first digital images were received from Earth observation satellites in the 1960s. However, since that time there have been some important and major changes for image processing driven by computer technological advances in hardware and software configurations.

The first images were received mainly by government bodies or satellite operators and were conse-quently processed solely by these organisations into hardcopy products before other users received them. This, together with the status of computer technology, being mainframe based in the 1960s, meant that the processing was carried out by specialist staff using tailor-made purpose-written software on dedicated computer platforms. Standard computer hardware was often modified to run one-off programs written by staff within an organisation and processing was not generally portable to other organisations or users.

There were no ready-made packages available for satellite image processing in the early 1970s. Instead people used to write their own routines using FORTRAN 77 or a similar programming language on mainframe computing systems. Purpose-built specialised image processing hardware systems were developed towards the end of the 1970s and were quite widely adopted by institutions in the early 1980s, often linked to host mainframe computers. Examples include the VAX 5750 running IIS (International Imaging Systems) processor, the GEMS image processor running on a PRIME computer host and the IGAS system run by BP until the early 1990s. Many of these systems were developed for a specific sector of users, e.g. geologists, environmental scientists or meteorologists.

However, in the weather forecasting community, it was soon realised that processing could not be done centrally by one organisation, particularly if regional and global forecasts were required. Networks were therefore built up, so that data could be sent to other regional forecasting offices and these offices were given duplicates of hardware and software in order to effect uniform processing across nations and forecasting regions.

The software was often tailored to provide only the image enhancements required for a particular application. This meant that if you were going to set up an image processing facility you would buy the system that had been developed for your application. Generalists tended to favour systems such as GEMS and IIS as these were more flexible in the application provision and their link-up possibilities to mainframes. In the mid-1970s and early 1980s several software companies developed specialised image processing computer units which could be linked up to standard mainframe computers. Examples include the American International Imaging Systems processors, the UK GEMS of Cambridge and the MicroBrian system developed in Australia. These systems had processing programs 'hardwired' into dedicated processing systems. However, they relied on a 'host' mainframe computer for data input/output and peripheral handling procedures.

The problem with these systems was that they were bulky, required air-conditioned environments and were expensive to purchase, run and maintain. This meant that relatively few organisations could afford to buy and justify running the commercially available systems, thus limiting the number of remotely sensed images that were utilised globally until the mid-1980s. Many people who would have used satellite images were unable to do so unless they paid for a consultant to process the data for them or hired the facilities from one of the few organisations such as the UK National Remote Sensing Centre in Farnborough. However, another revolution in computer technology improved access to image processing facilities in the mid-1980s. Computer desktop workstations were developed by companies such as SUN Microsystems, Digital and Silicon Graphics. At first these provided twice the processing power of the old mainframe systems for the same price, but as mass production developed and microchip technology advanced in the second half of the 1980s, workstations became more affordable and offered many times the processing power of their mainframe ancestors. Software companies began to write general-purpose, multi-application image processing software packages that were compatible with all the makes of workstation and the UNIX operating system at first, but later using other operating systems too. Some of the companies that had produced the 'hardwired' mainframe systems adapted and remodelled their software to be compatible with workstation systems, providing many 'off the shelf' ready-written software routines for users who had little or no knowledge of programing.

Soon after came the Graphical User Interface (GUI) (i.e. Windows interface with menu and mouse-driven routines), making the software more user-friendly and more accessible to people with no computer programing knowledge. This then stimulated new companies such as Floating Point Systems to come into the market and revitalised some of the old companies to revamp their software from mainframe hardwired routines to software packages supported on various workstation platforms.

Recently there has been a big drive in the production of PC (Personal Computer) versions of traditional packages supported by Pentium PCs and the NT operating system. In the mid-1990s image processing became even more affordable and accessible to users with the development of more powerful PC Pentium processing. Many software packages originally written for workstation computer platforms have been adapted to run on PCs with Windows 95 to 2000 interfaces. This has brought professional and sophisticated image manipulation to a cost of only a few thousand pounds/dollars and is thus affordable to many more users.

Current Hardware Platforms

The most commonly used hardware platforms for image processing of Earth observation images are SUN Microsystems, Silicon Graphics workstations, Digital workstations or PC Pentium computers. Basic configurations of these machines can be bought for under £10,000 sterling and suitable PC Pentiums for under £2,000 sterling. In addition to a hardware platform for processing, users of image processing software require various computer peripheral devices so that data can be loaded into the computer and final processing results can either be printed out or backed up onto magnetic storage media. Common data entry configurations include CD-ROM drives, exabyte or cartridge readers on workstations or CD-ROM and Zip or Jazz diskette drives on PCs. Data can be loaded onto PC from floppy disks, but the capacity of a standard High Density double-sided diskette is only 1.4 megabytes whereas a whole Landsat TM scene occupies 262 megabytes. Only a 512×512 subsection of four or five bands of Landsat TM imagery can therefore be stored on one standard diskette. This means that, whilst a diskette may be adequate for data entry and storage for someone learning how to use an image processing package, many professional users require greater magnetic data storage capacity, and thus use CD-ROM read/write drive optical disks, Jazz or Zip drives. Disk media are preferable to tape media, as files are directly accessible without the need for sequential searches and are generally more reliable and robust. More recently high-density diskettes have become available such as the LS120. Digitisers can also be used for data entry, particularly for GIS, map or field data entry. Digital photographic scanners are increasingly being used to make digital inputs from hardcopy images. Output media include plotters and colour printers, which vary from A5 to A0 size. There are so many different makes of peripherals on the market at present that it is impossible to mention any by name here. However, when users are setting up an image processing facility, they are advised to consult their computer hardware and software suppliers to obtain advice on the best configuration of peripherals for their particular requirements. Readers interested in purchasing their own image processing software should refer to Appendix D for names and addresses of suppliers to contact.

Current Software

All the major players in the Earth observation software providers' community have upgraded their software to have PC or workstation Graphical User Interfaces (GUI) over the past few years. Many have also adopted the now familiar Windows format with standardisation of icons and menu options, with a few characteristic variations. The major advantage of this is that the software is much easier to use and learn than earlier versions. It also means that once the rudiments and principles of image processing have been learned and practised on one software package, it is relatively easy to use this knowledge on another commercial software package, thus enabling transferable employment skills. Another advantage is that recent versions of image processing packages can more readily import or export GIS data formats (Graham and Gallion 1996). The move to more widely available and cheaper hardware has made image processing

more accessible to more people. This has created increased market demand and consequent economies of scale have driven down the cost of software itself (depending upon the package bought) from a few hundred to a few thousand pounds sterling.

The major disadvantage is that, in making software easier to use, the packages have become more difficult to tailor to specific applications. It is also more difficult and/or expensive for experienced image processing practitioners to access the source programming code to make modifications to create or tailor-make application-specific routines. Some software packages have macro languages and/or toolkits to set up user routines, but these are usually available only as additional modules at extra cost. Several leading commercially available image processing packages will now be briefly examined. However, this list is not exhaustive.

ERDAS Imagine

ERDAS Incorporated, a supplier of image processing software, was founded in 1978 as a commercial spin-off of research developed at Harvard University and the Georgia Institute of Technology. ERDAS Imagine was developed from a command-line software package called ERDAS 7.5. It operates on all major Unix workstations. It is now available to operate on PC with NT or Windows 95 and later operating systems. ERDAS Imagine is an easy-to-use, customisable software package which provides a full suite of image processing, photogrammetry, GIS analysis, database and visualisation tools. Its particular strengths are in land-use classification, ortho-modelling, making image maps and interfacing/data exchange with geographical information packages, having been developed closely in parallel with ESRI's ArcInfo software.

ERMapper

ERMapper was developed by the Perth-based Australian company Earth Resources Mapping in the early 1990s, initially for the geoexploration community of users and therefore is particularly suitable for enhancing geological features in satellite and airborne images. Like ERDAS Imagine, it is available for Unix and Windows 95 or later or Windows NT operating systems.

EASI/PACE

EASI/PACE has been developed and marketed by PCI, a Canadian company, and is less well known than ERMapper and ERDAS Imagine in the UK and continental Europe. However, it is a strong competitor in the market, providing all the standard image processing functions plus specialised advanced classification, and ortho-photo rectification functions. It was also the first commercial general-purpose image processing package to produce a specialised and comprehensive module for effective radar image processing. It operates on all major Unix workstations. On PC it runs under Windows 3.1, Windows 95/98, Windows NT and OS/2. It can also operate on VAX systems.

ENVI

ENVI is supplied by Floating Point Systems, and is based on the programming language IDL. Users can purchase ENVI as a 'runtime'-only licence or opt for more flexibility and purchase IDL as well so that they can write their own routines. ENVI is the first package to offer advanced spectral enhancement and manipulation capability so that users can perform hyperspectral modelling on multiband image datasets, such as CASI and the Airborne Thematic Mapper.

Dimple

Dimple is a Macintosh-based package, developed and supplied by Cherwell Scientific of Oxford, UK. It provides all the standard image processing functions and can support data in ERDAS and MicroBrian formats. Its main advantage is that it can operate on the Macintosh system, but is not very good at handling large datasets. A PC version of Dimple is currently under preparation.

TerraVue

This is a low-cost PC Windows package for image processing. It provides all the standard image pro-

cessing functions plus some more sophisticated filtering operations, atmospheric correction, mosaicing and rectification manipulations. It requires a 486 or more powerful processor.

TNTmips

TNTmips is part of a suite of packages that perform image processing (TNTmips), visualise and interpret data (TNTview) and publish spatial information (TNTatlas). These packages, developed by MicroImages Incorporated, Nebraska, USA, are very powerful, providing all the standard functions plus comprehensive stereo-viewing and DEM creation modelling with processing flexibility. TNTmips can be used on PC with Windows 3.1 or 95 or later or NT, Macintosh or on UNIX workstations. TNTmips is one of the older commercial packages that has been translated from command-line operation on a mainframe computer to Windows on a PC. This tends to make it harder to learn initially but, once mastered, it is very powerful package to use.

Idrisi

Idrisi was developed by Clark University, Worcester, USA, initially as a GIS teaching package, and it is still not strictly a commercial package, although many universities now use it as a more cost-effective alternative to the commercial workstation packages. Although it was originally written as a GIS package, it has been adopted for simple image processing because of its good range of standard basic image processing functions. It has been adopted by overseas aid organisations for installing cheap, robust and PC-based image processing and GIS capabilities in developing countries. It runs on PC platforms with 386 processing power and above, although the Windows version requires a 486 or higher specification PC.

IGIS (Integrated Geospatial Information System)

This system was jointly developed by Laser-Scan, NERC and the UK Defence Research Agency as the first workstation-based combined GIS and image processing package. It combines all the standard image processing functionality with some basic GIS functions such as digital terrain modelling, hill shading, plotting, digitising and editing, slope and aspect map generation and raster-to-vector conversion. It can be supported on both SUN and Digital workstations and uses Oracle or Ingres relational database management systems.

DRAGON

DRAGON was developed by Goldin–Rudahl and is an inexpensive PC-based image processing system, used in higher education and schools. It includes all the standard image processing functions and runs on PCs with 386 processing power and upwards. A limited version of DRAGON is included on the CD accompanying this volume and more details of its functionality are provided in the Practical Manual.

These packages are being updated and improved continuously, with most companies maintaining good communication with current users through telephone, e-mail helpdesks and regular Usergroup meetings. For up-to-date specifications and capabilities of particular packages readers are advised to contact the relevant suppliers and/or software houses to obtain their latest information and price lists (addresses can be found in Appendix D).

Choice of package will depend upon your application as most software packages can do all the standard image processing functions (such as display, geocorrection, simple image rectification, contrast stretching, statistics, simple band arithmetic functions and basic classification routines). However, some are more specialised in specific application areas than others are, which is often a reflection of the original reasons for their development.

Software Evaluation

Software evaluation is increasingly becoming an important process as the choice of image processing packages on the market increases, the relative costs are reduced and the flexibility of packages increases. Capital outlay on hardware and software for image processing has always been a fairly major decision for any organisation. This is changing with PC technology but it is still a major purchase to justify for an organisation or project using image processing for the first time. The following section provides some guidelines on how to evaluate software.

What is a software evaluation?

It is usually a carefully researched and prepared report and/or a demonstration of the capabilities of a candidate or several candidate software packages for a specific task, project or working environment.

When might you be asked to evaluate software?

The remote sensing specialist in a non-remote sensing organisation may be asked to specify suitable image processing equipment for the set-up of a new unit or facility, or an organisation may want to update its existing facility. A customer project requirement may need specialist image processing, e.g. hyperspectral image analysis, determining the purchase of a more specialised package for the task, or the customer may ask for advice on setting up new image processing facilities in their organisation. All software packages on the market today tend to offer a standard range of image processing functions. However, the features which will distinguish them from others will be in their:

- user-friendliness/training and the amount of training required before they can be properly used;
- cost;
- flexibility of data formats for input/output, hardware platforms and peripherals;
- post-sales support, frequency of software updates and licensing requirements;
- unique and defining functionality. For instance, they may have topographic/hyperspectral modelling routines, radar-processing capability, knowledge-based/neural network processing or better

interfaces with other non-image processing software packages, especially GIS and database packages or DTP (Desk Top Publishing).

There may be a trade-off between these, for example a loss in user-friendliness, increased cost or a compromise on standard image processing functions. Most software vendors have demonstration versions of their software which are either reduced functionality versions of the software or the full licence software is available for a limited time, typically one month. A fee may be charged for this access although it is often free. Other vendors prefer either to come and demonstrate the software to you or to let you use it under their supervision at a conference, exhibition or their office premises. If possible, try to obtain an evaluation or demonstration copy of the software on a temporary licence, otherwise you will not be able to do a thorough enough evaluation.

Developing a strategy for software evaluation

When developing a strategy for software evaluation you will need to take the following points into account.

- An uninterrupted time period is required.
- Ideally, use the same image data for all the packages that you are evaluating. Also, identical hardware platform configurations, operating systems and set-ups must be used for all the packages being evaluated.
- An evaluation strategy must be set up and applied under the same conditions to each package being evaluated.
- A standard method for recording the evaluation test results should be used, for instance a table or computer spreadsheet.
- Prepare samples of image datasets which encompass the types that you commonly use in your work and apply your evaluation processes to these. For example, some packages will handle small datasets efficiently but cannot handle large datasets very well.
- Plan the strategy by using a series of standard

image processing functions to test the software functionality and efficiency of use. These should test the following:

- Are all the standard image processing functions available? Image display, band selection/colour resolution of display, automatic and user-defined contrast stretches, geocorrection and image rectification, radiometric/atmospheric correction, image mosaicing, statistical analyses, range of both spectral and spatial image enhancement tools, data/pixel queries, classification including supported algorithms, GIS package compatibility, input/output and terrain analysis functionality.
- How long do processes take to run?
- How many temporary/processing files are produced on the way?
- Is the quality of the end result suitable for your particular application?
- Does the software provide error trapping in processing or in inaccurate data input?
- Does the software provide accuracy testing for processes such as geocorrection/classification and how thorough is this? Is it efficient or time consuming?

Other features that should be compared are listed below:

- Cost, supplier location, after-sales service – can you purchase a licence for multiple copies of the software? Is the software available through any educational discount schemes such as CHEST (Combined Higher Education Software Team)?
- Computer platforms/operating systems that can support it.
- User-friendliness – can the software be used with little or no knowledge?
- Tutorials – does the software have training tutorials and are they easy to follow? Do they tell you how to use all the standard image processing functions?
- Does the software have on-line help or manuals and are these well written and easy to understand?

- How fast is the speed of processing on complex tasks such as classification, geocorrection and ortho-correction?
- Does the package have flexibility for host platforms and operating systems (PC and work-station)?
- Is data import/export easy and are the most common data formats and types supported (hardcopy as well as softcopy)? Is the software able to process large datasets? Is it easy to integrate data from other packages, particularly all the commonly used GIS packages?
- Is the software compatible with peripherals: print-ers/plotters/digitisers/scanners?
- Can you write your own routines and readily inte-grate them into the software? Are there any tools to help you do this?
- Does the package have non-standard functionality: for example, ortho-correction/3D visualisation, handling GPS data, radar image processing, hyperspectral processing, CAD/desktop software interfaces?

Conclusions

Choice of hardware and software for image processing is still a fairly major purchase for the user and requires some thought and consideration, particularly if specialised applications of Earth observation are required. Image processing software and hardware supply is a competitive marketplace, providing the user with a good range of products; it is therefore important to contact suppliers, ask advice and obtain demonstration versions before purchasing. If only standard basic image processing procedures are required, then all the main packages mentioned in this section will provide suitable functionality. How-ever, for specialist applications some packages are bet-ter adapted than others. For example, for interfacing with ArcInfo modules and ortho-modelling ERDAS Imagine is best, but for image enhancements for geo-exploration, ERMapper is the most suitable. DRAGON is an excellent system for 'classroom teaching' of digital image processing because it is rel-atively cheap to purchase multiple licences and it is

fairly simple to operate so that students can learn to use it quickly. Idrisi provides an easily used Windows PC-based package which requires little technical sup-port on low-cost hardware. However, new updated and improved versions of all the packages are brought out at frequent intervals, often remedying former weaknesses and providing modules to cater for former omissions. For latest developments in software and hardware products readers are advised to contact the software suppliers directly and visit the commercial exhibitions at annual remote sensing and GIS conferences.

3.12 CHAPTER SUMMARY

- A number of digital processing enhancement pro-cedures can be applied to image data in order to highlight specific features. The optimum tech-nique depends on the research interests of the operator.
- Most images require some form of contrast stretching before they can be examined because the original DNs are often concentrated within a small range. Simple linear stretches may be applied or non-linear stretches such as histogram equalisation, logarithmic or power-law stretches, which preferentially stretch specific parts of the DN range. The mapping of one range (unstretched) to another (stretched) can be accom-plished by means of a look-up table (LUT).
- Ratio images are formed by dividing the digital number in one band by the corresponding DN of another band for every pixel in the scene. The infrared/red ratios are particularly important in vegetation studies whilst other ratios have been employed in geological investigations.
- An image, which can be thought of as a combination of long-, medium- and short-wavelength features, can be filtered in order to accentuate a specific wave-length. Low-pass filters have a smoothing effect whereas high-pass filters emphasise edges. The filtering may be performed in the spatial domain by convolution filtering or in the frequency domain by the use of Fourier transforms.

- A colour image, instead of being viewed in a red, green, blue context, may be thought of in terms of intensity (brightness), hue (a measure of the colour) and saturation (a measure of the colour purity). Transforming the data into an Intensity, Hue, Saturation image may accentuate specific features.

- A principal components transform can be applied to image data in order to produce 'new' principal component images which are non-correlated. Typically the first PC contains over 80 per cent of the data variance, with progressively lower variance in the higher components. Colour images produced from principal components are generally more colourful than images produced from the original data.

- Images produced from data obtained from different remote sensing systems can often provide more information than images constructed from a single system. Data can be combined in a number of ways, for example by projecting images from different systems in different colours or by means of an IHS transform. Draping the image data over a digital elevation model can generate perspective views of the terrain.

- Classification allows an area to be divided up into a number of themes based on the pixel values in the dataset. Themes may not represent single surface classes but only show areas which have similar spectral characteristics. Supervised classification involves the image analyst using independent knowledge to define training areas which do represent specific surface classes and using the statistical characteristics of the training areas in an attempt to assign unknown pixels to specific surface classes.

SELF-ASSESSMENT TEST

1   An unstretched image has a DN range from 55 to x which is subsequently linearly stretched such that 55 is assigned a value of 0 and x a value of 255. If 75 in the input image is allocated a DN of 68 in the stretched output image, what is the value of x?

2   How many simple ratios (one band divided by another where A/B = B/A) can be created for an 8-band system? If these ratios are to be combined into 3-band false colour composites, how many different ones can be produced?

3   The DN histogram for an image is displayed in Figure 3.32. Calculate the output DN range for the input range 45–60 when:
   (a) the full input DN range (20–150) is stretched to 0–255;
   (b) the range 20–80 is stretched to 0–255;

4   What is an edge on a remotely sensed image?

5   Describe what the LUTs shown in Figure 3.33 will achieve.

6   The DNs for an image are shown in Figure 3.34 and they indicate the presence of a vertical edge (shown shaded) where DN values are lower than the background. Apply the given low-pass filter to this data and discuss how the DNs now vary across the edge. Also, apply the given high-pass filter, add the resultant image back to the original and discuss how the DNs change across the edge for this image.

Low-pass filter			High-pass filter		
1	1	1	-1	-1	-1
1	1	1	-1	8	-1
1	1	1	-1	-1	-1

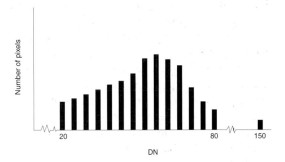

Figure 3.32 Histogram to be used with self-assessment question 3.

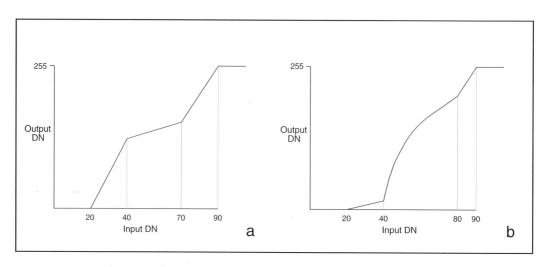

Figure 3.33 LUTs to be used with self-assessment question 5.

Original DN

19	20	18	2	19	17	18
22	24	20	5	24	19	14
27	32	23	1	18	29	19
28	36	21	4	21	30	16
27	44	30	3	27	23	10
36	50	44	2	44	28	13
33	46	43	3	36	44	25

Figure 3.34 Image matrix to be used with self-assessment question 6.

FURTHER READING

Avery, T. E. and Berlin, G. L. (1992) *Fundamentals of Remote Sensing and Airphoto Interpretation*, 5th edition, Englewood Cliffs, NJ: Prentice-Hall (Chapter 15).

Campbell, J. B. (1996) *Introduction to Remote Sensing*, London: Taylor and Francis (Chapter 11).

Cracknell, A. P. and Hayes, L. W. B. (1991) *Introduction to Remote Sensing*, London: Taylor and Francis (Chapter 9).

Jensen, J. R. (1996) *Introductory Digital Image Processing: a remote sensing perspective*, Englewood Cliffs, NJ: Prentice-Hall (Chapters 7, 8 and 9).

Mather, P. M. (1987) *Computer Processing of Remotely-Sensed Images*, Chichester: John Wiley and Sons (Chapters 5, 6, 7 and 8).

Sabins, F. F. (1997) *Remote Sensing: principles and interpretation*, New York: W. H. Freeman and Company (Chapter 8).

4

INTRODUCTION TO ENVIRONMENTAL MONITORING TECHNIQUES FROM SATELLITE DATA

4.1 INTRODUCTION

An application in remote sensing is the practical use to which a series of aerial or satellite images are put. The application of remote sensing or Earth observation techniques to atmospheric, Earth and environmental sciences can vary according to the final user's requirements.

1 Baseline environmental or resource assessment: remote sensing is used for making an assessment of different features at one point in time. This can be carried out as an isolated study such as assessing an area for its potential for mineral exploration.
2 Change detection: baseline resource assessments can be made at several discrete points in time, for example at annual intervals through a ten- or twenty-year period to assess landscape change arising from housing development at the urban/rural fringe of a city.
3 Environmental monitoring: daily and hourly satellite data can be used for continuous/real-time assessment of a dynamic environmental situation, for example development of a tropical storm or a coastal pollution event.
4 Mapping: remotely sensed data are particularly suitable for mapping environmental phenomena at a number of different scales.

Each of these modes of using remote sensing has specific requirements that make a particular application viable. These are related to the scale or resolution of the dataset, and its frequency of delivery to the user. Often environmental projects that use remote sensing will employ more than one of these modes; see case studies in Chapter 5.

Environmental monitoring, in contrast to environmental assessment, is concerned with not just a static assessment/observation of an environmental parameter at one point in time. Rather, it means being able to update information and observations of these parameters regularly in order to monitor dynamic processes. Therefore, environmental monitoring requires a regular supply of data on the environmental parameters being observed/assessed. The first applications in environmental monitoring using satellite image technology were in the field of meteorology and weather forecasting. Meteorological applications are considered in the first part of this chapter. This is followed by a consideration of how remotely sensed imagery can be used for vegetation and land-cover assessment in the second part of the chapter. Chapter 5 presents some case studies to demonstrate how remote sensing can be used in different environmental assessment and mapping projects.

Rates of Change of Environmental Phenomena/Frequency of Satellite Data Supply

Different environmental applications require different frequencies of information updates for monitoring to be effective. Figure 4.1 shows typical applications plotted with respect to the frequency with which they require information updates (temporal resolution) and the spatial detail of information that is required (spatial resolution). Environmental phenomena such as weather systems, natural hazard extreme events, pollution, or oceanographic events are very dynamic and rapidly develop over minutes and hours. Therefore for satellite data to be useful in their analysis, imaging frequency and data delivery has to be at least several times a day. At present only low spatial resolution meteorological satellite data can meet this need. Other applications such as crop monitoring require better spatial detail but rates of change occur only over a matter of weeks and therefore image updates need not be more frequent than weekly or monthly.

The main Earth observation satellite/sensor systems plotted against their temporal and spatial resolutions are shown in Figure 4.2. When they are compared with Figure 4.1, the frequency and scale requirements of environmental applications can be matched to the current provision by environmental satellite and sensor systems. This is very useful when one is planning the use of satellite imagery to monitor an environmental phenomenon. The cost of a square kilometre of data varies greatly between different satellite systems (Figure 4.2). There are examples of applications from different parts of these diagrams in this chapter and also in Chapter 5. We will return to the issue of making appropriate choices of satellite/airborne imagery for environmental applications at the end of this chapter.

Satellite Data Sources for Environmental Monitoring

The key satellite and sensor systems used for environmental monitoring are the meteorological satellites – GOES (including Meteosat) and NOAA AVHRR – as these satellites produce daily imagery of most of the Earth's surface. These satellites are used because of their frequent data collection, the size of the area that is imaged and because their imagery is available free to anyone with a suitable satellite receiver. Some environmental phenomena do not need such frequent updates of imagery but do need more spatial detail in order to be assessed. Therefore Landsat Thematic Mapper, Landsat MSS and SPOT multispectral (XS) data can also be useful for environmental monitoring. Since many environmental phenomena are likely to occur beneath cloud cover or at night, radar imagery such as that obtained from ERS or RADARSAT can often provide invaluable data. This will become increasingly

useful as scientists better understand the interpretation of radar data. This book concentrates on applications of non-radar data to environmental monitoring although the companion volume contains examples of radar imagery and information on the theoretical concepts and use of radar imagery.

PART 1: METEOROLOGY AND WEATHER FORECASTING

4.2 INTRODUCTION

Meteorology and climatology are considered first as they can be fundamental to understanding other environmental parameters. Many monitoring applications have been developed by the use of meteorological satellite measurements of the rainfall regimes in an environment in order to assess the status of vegetation/crops, risk of drought, bush fire or insect infestation. Initially, examples of methods for convective rainfall assessment over temperate, tropical and equatorial regions will be described as these are relatively easily understood. Weather system analysis over temperate regions will then be considered, and finally satellite data contribution to national weather forecasting and severe storm assessment.

Assessment of the weather is fundamental to the welfare of human populations world-wide. Rainfall is the most variable meteorological parameter and is a key limiting factor to crop and rangeland production. In excess, it can also have devastating effects such as floods and soil erosion. Increasingly, therefore, dense and accurate rainfall observations are required to meet the needs of growing human populations and their demands on marginal lands. However, over the last two decades the availability and reliability of rainfall data from conventional rain-gauge networks, particularly in tropical and subtropical regions, has declined. In addition, rain-gauge datasets record at point locations and do not provide a spatially continuous record of rainfall. Large areas of the world are devoid of rain-gauges, particularly in remote areas, and spatial interpolation

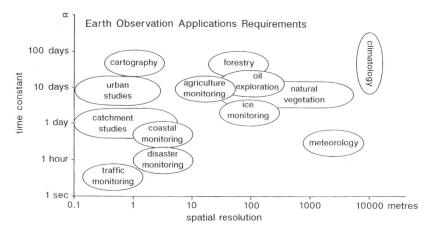

Figure 4.1 Temporal and spatial resolutions required for environmental applications.

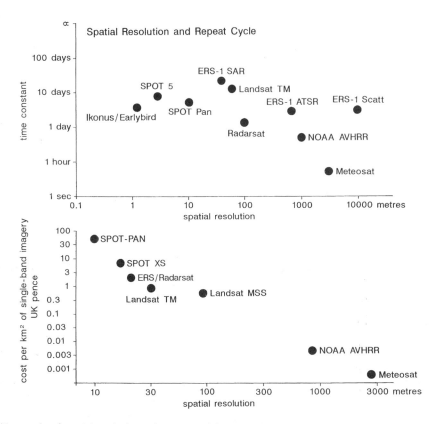

Figure 4.2 Temporal and spatial resolutions of current and future satellite imaging systems. Lower figure shows approximate cost of the data.

is difficult. Even where spatial interpolation can be performed, spatial and temporal variability of rainfall can give rise to serious errors or over-generalisations.

Meteorological satellite data provide wide-area continuous coverage at frequent enough intervals (30 minutes to every 6 hours) to enable rainfall monitoring or, at least, rain-cloud assessment in areas deficient of rain-gauge cover. Satellite rainfall estimates complement rain-gauge measurements by giving a better idea of the spatial distribution and variability of rainfall and by filling the gaps between rain-gauge locations.

Satellite data can provide information on cloud cover and rainfall in the visible, thermal infrared and microwave regions of the spectrum. The early techniques used visible and infrared imagery analysis but recent research is now developing reliable techniques for monitoring rainfall from passive and active microwave data. None of the techniques gives direct measurements of rainfall. Estimations are derived from either cloud mapping and analysis (nephanalysis) or in the case of microwave instruments by electromagnetic radiation waves having physical interaction with the rain droplets (i.e. scattering effects). Passive microwave frequencies most suitable for rainfall detection are 19.35 Gigahertz (GHz), 37 GHz and 85.5 GHz, but there are some difficulties with discriminating between rainfall and background moisture over water bodies and wet surfaces. This is an area of active research and development, and Lovejoy (1981), Bailey *et al.* (1986) and Kidd (1997) give detailed information on rainfall monitoring using passive microwave imagery and ground active radar rainfall measurement. The interested reader should refer to Appendix F where a full list of references is provided.

4.3 NEPHANALYSIS

Nephanalysis (the construction of cloud charts/maps) and cloud-cover mapping existed before the satellite era, the first serious analyses being carried out by ground and aerial surveys during the 1940s. However, since the general availability of satellite data in the 1960s, nephanalysis has been performed almost solely on meteorological satellite imagery.

In the visible part of the electromagnetic spectrum, high reflectivity of cloud indicates high water content and thick cloud cover, and therefore a higher probability of rain. However, cloud temperature, drop size distribution, concentration and evaporation are also important factors in rain-cloud determination (Barrett 1979).

In the thermal infrared part of the spectrum, low cloud top temperatures indicate high cloud tops, higher in the atmosphere. Generally, the higher a cloud is, the thicker it is and therefore it is usually a convective rain-producing cloud. However, not all high cold cloud is rain-cloud: jetstream cloud is high ice cloud but it does not produce rain, is not very thick, and has predictable locations over the globe. In areas of warm, low-level and stratiform rain-producing cloud, it is often more difficult to discriminate rain-cloud from non-rain-producing cloud.

The weather satellite programmes of NASA and NOAA have defined six criteria for successful cloud identification on satellite imagery: brightness, texture, size, shape, organisation and shadow effects (Conover 1962, 1963; Lee and Taggert 1969). Nephanalysis techniques seek to define different cloud types and cloud layer level from features definable on satellite visible and infrared data.

These characteristics, however, utilise subjective human visual interpretation, relying on the skill and experience of interpreters carrying out manual interpretations. The computer era and digital imagery created the possibility of automating the nephanalysis procedure so that the process becomes more objective and consistent, based on quantitative or objective analysis of greyscale values associated with cloud cover masses. In the thermal infrared, greyscale values can be calibrated according to cloud temperatures so that areas of different cloud temperatures can be mapped, as shown in Table 4.1. Nephanalysis is also used to identify weather systems such as frontal or depression systems, convective cloud development and for defining cloud climatologies of several regions of the world.

Until 1979 nephanalyses tended to be satellite cloud type distribution studies carried out by researchers for the NASA and NOAA Weather Satellite Programmes, and there had been no systematic studies of cloud type distributions and frequencies over other parts of the world. In 1979 Barrett and Grant carried out statistical and climatological comparison studies on 12 months of Landsat MSS data with contemporaneous ground observation data from reporting stations in the UK. This study proved useful, both for forecasting and in the understanding of local variations in climatic characteristics over the British Isles.

Satellite Cloud Classification Techniques

Most satellite cloud type classification techniques are known as supervised, with the cloud type groups pre-defined before the analysis. However, the following unsupervised techniques are also available (Hendersen-Sellers 1984):

1 Clustering and statistical techniques
2 Spectral/temperature difference techniques
3 Thresholding/density-slicing techniques
4 Plot separation techniques

Clustering and statistical techniques use image features more complex than temperature or grey-value thresholds in order to attempt to find natural groupings of pixels in an image. This procedure can result in a classification becoming too detailed by

subdividing logical classes. The other unsupervised techniques of classification are based on well-established temperature/brightness relationships translated into greyscale values on satellite imagery. These techniques are more commonly used as they employ image processing functions which are readily available in current image processing software packages and are based on the assumption that a brightness (greyscale) response in the visible or infrared on an image pixel represents a single temperature value. In the case of cloud cover, brighter, whiter greyscale values represent colder temperatures such as those that occur in the tops of high-level precipitating cloud (Bunting and Hardy 1984; Lovejoy and Austin 1979a and b). The main problems with such unsupervised schemes are:

1 There is no standard cloud classification scheme and therefore it is difficult to compare results.
2 There is disparity between ground-station data (viewed from below) and satellite data (viewed from above).
3 Classification algorithms, in general, have undergone very little testing and many are scene or region specific, therefore not universally applicable, particularly to areas where different climatic regimes operate from those on which the algorithm was originally devised.
4 Original satellite, calibration and verification datasets may not satisfy classical statistical assumptions.

Table 4.1 Relationship between greyscale and thermal infrared cloud temperature for Meteosat imagery over West Africa

	Altitude (m)	Pressure level (mb)	Temperature (K)	Greyscale brightness value on Meteosat IR images
Ground surface	0–3,000		> 289.5	144
Low level	< 3,000	> 700	> 285	153
Medium level	3,000–9,162	> 300	284–242	135
High level	> 9,162	< 100	< 241	63

The Development of Reliable Cloud Classification Techniques

A technique using the temperature difference between the ground/ocean surface and the radiating cloud surface to determine the size and type of clouds present in an image was developed by Rao (1970). This method needs temperature data from the ground or ocean surface and the radiating cloud surface in addition to lapse rate data and cloud height/temperature curves for the area being studied. These latter two data items are not always available for certain areas of the world. Since 1970 some researchers have refined cloud-cover mapping by using differences in cloud temperatures in the thermal infrared to determine height, amount and type of cloud. Others have regressed ground measurements with the results of a cloud-cover estimation from visible and thermal infrared images to derive 'guess fields' at time intervals from 5 to 15 days. These methods could be successfully combined with conventional rain-gauge data to provide 'optimal satellite estimates' of rainfall at locations where conventional rainfall data are not available (Reynolds and Vonder Haar 1977; Creutin *et al.* 1986).

Wielicki and Welch (1986) used Landsat MSS digital data to define cumulus cloud properties, by utilising a thresholding technique. The greater spatial resolution enabled them to define cloud fraction and reflectance as a function of cloud size and amount, and make reasonably accurate estimates of cumulus cloud cover, using greyscale thresholding. However, as yet the temporal frequency of Landsat and other sources of medium/high spatial resolution data is not great enough for current synoptic forecasting requirements.

Several researchers have developed bi- and tri-spectral (visible, near and thermal infrared) plot separation techniques using both GOES and NOAA AVHRR imagery. These plots include data from image wavebands at 3.7 and 11 µm. The methods are built on the basic principle that different cloud types and land or water surfaces have different radiational properties in different parts of the electromagnetic spectrum. This principle can be successfully combined with statistical clustering and plot techniques to create powerful cloud-classifying algorithms using a multidimensional feature space (Pairman and Kittler 1986; Desbois *et al.* 1982). Cloud-cover types are separated from snow and clear skies, using the digital greyscale spatial and spectral characteristics of GOES visible, near-infrared and thermal infrared images. The values are plotted in visible/near-infrared and/or thermal infrared frequency distributions to define discrete characteristics of each class type, to guide future classifications. The results are verified using synoptic ground-station data (Tsonis 1984; Liljas 1982, 1987).

Comparison between satellite and ground data improves cloud assessments, and satellite cloud estimation techniques provide a quantitative input to an otherwise subjective conventional method of recording cloud type and cover. The US Naval Postgraduate School has produced an automated cloud and precipitation model with the same objective as the multispectral cloud classification technique developed by Liljas (1982). It combines elements of several researchers' work (Barrett and Harris 1977; Reynolds and Vonder Haar 1997) to enable the separation of different types of cloud and in most incidences snow, sunglint, mist and fog. Up to 16 different classes can be identified in each daytime image. Processing night-time imagery was based on data from channels 3 and 4, calibrated into radiance units and then expressed as brightness temperatures. The simple box classification used by Liljas gives a fast, easily understood, automatic method of classifying cloud imagery. It has the flexibility to enable changing thresholds for different seasonal or synoptic situations, although some cloud types overlap in the classification (Figure 4.3).

Cloud and rainfall classification with Meteosat imagery is sometimes difficult because of poor discrimination between cloud and ocean in tropical and equatorial regions, where there is relatively warmer low cloud or fog over the ocean, or between clouds of a similar temperature or cold water upwelling from the depths of the ocean. This is a particular problem off the coast of West Africa because of the permanent presence of the cold Canary ocean current. Kelly

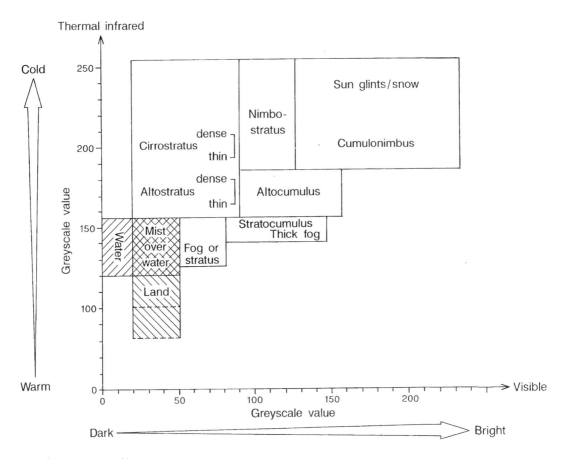

Figure 4.3 Clustering of bi-spectral plots, adapted from Liljas (1982).

(1985) found a solution to this problem by subtracting AVHRR channel 4 (11μm) from channel 3 (3.7 μm). Sea-surface temperature is reduced in comparison to the cloud and fog signal, so that clouds near very cold water can be separated much more effectively.

4.4 RAINFALL ESTIMATION

The basic assumption that underlies all rainfall estimation techniques which utilise visible and infrared satellite imagery is that precipitating clouds can be distinguished from non-precipitating clouds since

the former are usually thick (producing a high (bright) visible response) and/or tall (having cold cloud top temperatures and therefore producing a high infrared response). Since the late 1960s, a large number of rainfall estimation techniques have been developed using thermal infrared and visible data, in conjunction with weather data and/or climatological records for calibration and verification. Many of these have been developed for particular needs in specific areas and are not globally applicable. Atlas and Thiele (1981) and Barrett and Martin (1981) have provided syntheses of the numerous rainfall techniques and related cloud studies that had been developed during the first 20 years of operating meteorological satellites. Since then, new techniques have been developed and the current wide range of satellite-based rainfall estimation techniques using visible and/or infrared data may be summarised as shown in Table 4.2 (adapted and updated from WMO (1986). Most of the techniques are referred to by the names of their originators, and many have variations not included in the table.

Satellite-based rainfall estimates can be obtained using the techniques in Table 4.2 at an hourly temporal resolution and at about 5 km spatial resolution. These resolutions should be sufficient in the majority of cases since most precipitation occurs in storms that are 20–2,000 km in the horizontal dimension (Houze and Hobbs 1982). Smaller rain showers do not contribute significantly to precipitation amounts. There are many variations of the three main categories of rainfall estimation techniques (discussed below). Some have been designed for specific climatic regimes, others for short- or long-period rain events. Readers are directed to the reference list for specific information on these techniques. The following section gives a few examples of important methods in each of the main categories.

Over the last two decades investigations have been made into the use of the passive microwave data from experimental satellite missions. Passive microwave-based techniques have the advantage of being more physically direct than visible and infrared measurement as they involve absorption and scattering interaction between rain droplets and microwave radiation in the atmosphere, but sensor imaging is currently too infrequent for reliable forecasting or climatological monitoring. For further information on these techniques the reader is directed to Bailey *et al.* (1986) and Kidd (1997). Weather radar also operates in the microwave part of the spectrum, but these techniques are generally ground-based systems and are used in conjunction with rain-gauge data to calibrate and verify results from satellite rainfall estimation and forecasting techniques. However, most of the reliable operational methods are based on visible and/or infrared imagery, and this review will concentrate on these methods. At present the satellite rainfall monitoring methods best suited to operational use are those which integrate conventional rain-gauge, ground radar (when available) and synoptic station records with satellite-derived rainfall estimates or analyses. These four data sources both complement and enhance each other. Rain-gauges provide accurate point location rainfall amounts continuously through time, ground radar gives frequent instantaneous rain rates, and station records produce a complete spatial representation of rainfall patterns in an estimated form at regular, but separate, instants in time. Satellite data are thus particularly useful over areas where the distribution density of rain-gauges is low, measurements unreliable, and ground radar non-existent, for example over most major mountain regions and extensive humid regions of the tropics and in arid and semi-arid regions. These regions are typified by convective rainfall events which show extreme variations in temporal and spatial distribution. In most regions of the world the spatial resolution of the satellite sensors is finer than the spatial distribution of existing rain-gauge networks and so can offer a more detailed estimated representation of rainfall. Thus, operationally, satellite data are used to fill in the gaps between the locations of conventional data collection points. Geostationary meteorological satellites, for example Meteosat, provide the best source of satellite imagery for these purposes as the data are spatially continuous, temporally highly repetitive and relatively inexpensive to receive and process. In the United States or northwest Europe, where ground-based radar provides an additional source of conventional data, these are often incorporated into the algorithms to increase estimation accuracies.

Satellite rainfall estimation techniques may be divided into three major types according to their degree of automation (Moses and Barrett 1986):

1 Manual
2 Interactive
3 Objective (automatic).

These have been developed for the estimation of rainfall at a number of temporal and spatial scales,

Table 4.2 Summary of satellite rainfall estimation techniques

Method	Date of creation	Author/s	Bands used	Spatial resolution	Global climate zone	Manual Automatic Interactive	Timescale
ADMIT	1986	Barrett & D'Souza	VIS & IR	1–5 km	Equatorial Africa	A	1, 10, 30 days
BIAS	1984–5 ref. 1986	Moses & Barrett	VIS & IR	1–5 km	USA, Middle East, Africa	I	1, 6, 12, 24 hours
BOCHUM	1982	Kruger et al.	IR	5–25 km	Germany	A	0.5–1 hour
Bristol	1970–5	Barrett	VIS & IR	5–25 km	NW Africa, Oman	M	1, 6, 12, 24 hours
CEAS	1977	LeComte	VIS & IR	5–25 km	Sahel Africa	I	10 & 30 days
CIST	1981	Griffith et al.	IR	1–5 km	S. Florida GATE	M & A	0.5–1 hour
CROPCAST	1985	Heitkemper et al.	VIS & IR	1–5 km	USA	I	Daily
FAO	1981 1986	Van Dijk Hielkema	VIS & IR	25 km +		I	10 & 30 days
FRONTIERS	1979	Browning (UK Met. Office)	VIS & IR Ground RADAR		UK, W. Europe Atlantic	I I	0.5–1 hour
GPI	1979 updated 1983	Arkin et al.	IR	25 km +	GATE area, USA Tropics/mid-latitudes	A	10, 30 days
GRID HISTORY	1985 ref. 1986	Martin & Howland	VIS & IR	5–25 km	GATE area, S. China,	I	Daily
HRC	1976, updated 1981	Kilonsky & Ramage Garcia	VIS	25 km +	Tropical Pacific, Atlantic	M	10 & 30 days
MELBOURNE	1985	DelBeato & Barrell	IR	1–5 km	SE Australia	A	Daily
MICHIGAN	1983	Cutrim			Amazonia	I	Daily
MONTREAL	1979 ref. 1979a and b	Lovejoy & Austin	VIS & IR	1–5 km	mid-latitudes	A	0.5–1 hour
NASA/MARSHALL	1985	Robertson	IR	1–5 km	S. Pacific	A	0.5–1 hour
NASA/GLAS	1984	Negri et al.	IR	1–5 km	USA	A	0.5–1 hour
NESDIS	1977 1984	Scofield & Oliver Clark & Borneman	VIS & IR	1–5 km			0.5–1 hour

Table 4.2 Continued

Method	Date of creation	Author/s	Bands used	Spatial resolution	Global climate zone	Manual Automatic Interactive	Timescale
NESS	1979	Whitney & Herman	IR	5–25 km		A	1–6 hours
Nimrod	1995 ref. 1997	UK Met. Office	VIS, IR and ground RADAR	25 km +	UK, W. Europe Atlantic	A	0.5–1 hour
PERMIT	1986	Barrett D'Souza & Power	IR	1–5 km	Equatorial Africa	A	1, 10 & 30 days
RAINSAT	1982 & updated 1986	Bellon et al.	VIS & IR		N. America	A	0.5–1 hour
TAMSAT	1985 ref. 1987	Milford & Dugdale	IR	1–5 km	Africa	A	1, 10, 30 days

Methods Known by their Authors

Author/s	Date of creation	Bands used	Spatial resolution	Global climate zone	Manual Automatic Interactive	Timescale
Neil	1984, updated 1987	VIS & IR	1–5 km	S. Pacific	I	0.5–1 hour
Weiss–Smith	1987	IR	5–25 km	N. Dakota USA	A	0.5–1 hour
Inoue	1987		1–5 km		A	Daily
Doneaud et al.	1984 & updated 1987	IR	1–5 km	USA	A	Daily
Follansbee et al.	1975	VIS	5–25 km	USA	M	Daily
Creutin et al.	1986	VIS & IR	5–25 km	Middle East	M	1–10 days
Fenner	1982	IR	5–25 km	Europe	I	1–6 hours
Motell–Weare	1987	IR	5–25 km	Tropical Pacific	A	10, 30 days
Tsonis–Isaac	1985	VIS & IR	1–5 km	Mid-latitudes	A	0.5–1 hour
Wu et al.	1985	VIS & IR	1–5 km	Florida/Mexico	A	0.5–1 hour
Lee et al.	1985	VIS & IR	1–5 km	Florida/Mexico	A	0.5–1 hour

Table 4.2 Continued

Notes:

1 As several methods were created some years before they were documented, therefore the date of their *creation* is often not associated with a reference. Where a later reference is known, this has been denoted on the table by the abbreviation 'ref.' preceding the date of reference in the scientific literature. The word 'updated' preceding a date indicates that the method has been updated or improved on the date shown.

2 Full names or details of abbreviated methods:

ADMIT	Agricultural Drought Monitoring Integrative Technique
Bristol	Bristol Method
BIAS	Bristol/NOAA Iterative Scheme
CEAS	Center for Environmental Assessment Services
CIST	Conservative Infared Satellite Technique
CROPCAST	US Agricultural Forecasting Service
FAO	UN Food and Agriculture Organisation
FRONTIERS	Forecasting Rain Optimised using New Techniques of Interactively Enhanced Radar and Satellite
GPI	GOES Precipitation Index
HRC	Highly Reflective Cloud
IIC	Intensity Index of Convection
NESDIS	NESDIS Operational Precipitation Technique
Nimrod	UK Met. Office forecasting technique from 1995
PERMIT	Polar-orbiter Effective Rainfall Monitoring Integrative Technique
RAINSAT	Canadian Weather Forecasting Technique
TAMSAT	Tropical Agricultural Meteorology using SATellites
UC-SD	University of California, San Diego technique

and have been applied in many different areas and situations, with varying degrees of success. Many of the satellite rainfall estimation techniques developed to date have been tailored for particular needs in particular areas, and they include several different methods of deriving rainfall estimates from satellite sensors.

1 Manual Techniques

These generally comprise the old pioneering techniques developed during the 1970s and early 1980s when digital satellite images were not widely available. They are based on the analysis skills of a spe-

cialist interpreter, for instance to identify and evaluate convective clouds and cloud systems capable of prompting extreme rainfall events and their associated hazards (Scofield 1984). An example is the Bristol Method.

2 Interactive (Semi-Automatic) Techniques

These techniques involve stages of analysis which are pre-programmed on computer and do not require human intervention, combined with other stages requiring a skilled analyst to enter data or to carry out analyses on the image processing. They were devel-

Manual Technique (the Bristol Method)

This technique is based on expert visual interpretation of gridded hardcopy satellite imagery (mostly infrared), calibrated by climatology, topography and also by meteorology whenever synoptic data are available. The Bristol technique was developed by Dr Eric Barrett of the University of Bristol and NOAA and is particularly suitable for assessing rainfall from middle-level stratiform cloud in mid-latitudes.

Rainfall estimates (typically for 6-, 12- or 24-hour periods) are made indirectly from assessments of cloud type and area, weighted by synoptic weather reports and terrain influences (Barrett 1970, 1973). The relationship invoked by the Bristol Method assumes that:

$$R = f (Ca, Ct, Mc, Sw)$$

where accumulated rainfall R is a function of cloud area (Ca), in grid squares of selected size, cloud type (Ct), combined influences of climatology and terrain (Mc), and synoptic weather (Sw) by which raw estimates of rainfall may be adjusted up or down from the climatological norms as the immediate weather circumstances demand. The 'cloud indices' formed from (Ct × Ca) are translated into rainfall estimates through local region-specific regressions. For full details of the manual Bristol Method the reader is directed to Barrett (1979). This technique was most extensively tested over northwest Africa in the late 1970s to help demonstrate the value of satellite remote sensing in the identification of areas liable to desert locust population upsurges.

With the advent of the AgRISTARS (Agriculture and Resource Inventory Surveys Through Aerospace Remote Sensing) Program in the USA, it became possible to develop an interactive version of the Bristol Method, the Bristol NOAA InterActive Scheme (BIAS), for use on an interactive image processing system, which is a later developed computerised version of the Bristol Method.

oped in the mid to late 1980s and are often computer versions of the original manual techniques, for example the Bristol–NOAA Interactive Scheme (Barrett 1984). Interactive techniques are still employed in situations where meteorological variables are difficult to program automatically, for example in mid-latitude low and mid-level stratiform cloud situations (Bonifacio 1991). In general, interactive techniques have provided a means for combining human interpretational skills with the image enhancement and computational powers of computers. However, in many situations analysts may not always be available or the interactive interpretations are time consuming and can vary quite considerably. Much research has therefore been carried out on the development of fully automatic techniques which would produce faster, more objective rainfall estimates.

3 Objective (Automatic) Techniques

These techniques involve classifications based on some previously defined cloud and rainfall imagery characteristics incorporated into software, so that a classification can be carried out independently of any human intervention once digital imagery and any ancillary data have been introduced. These are used for the regular assessment of rain-cloud and rainfall over large continental areas and are particularly suited to convective rainfall arising from convective cloud systems with short life cycles. This type of rainfall requires high-frequency satellite data. Key methods of this type include the Scofield and Oliver Convective Precipitation Technique (1977), PERMIT and ADMIT (developed at the University of Bristol) and the Cold Cloud Statistics Technique from

An Example of an Interactive Technique (Bristol/NOAA InterActive Scheme, BIAS)

This technique is an improvement of the Bristol Method and is based on the interactive assessment of cloud area and cloud type, and the translation of these assessments into 'cloud indices' according to a menu. These cloud indices are in turn translated into rainfall estimates through global regression look-up curves. Initial estimates may be made for 6-, 12- or 24-hour periods, depending on the availability of images, and these initial estimates may be adjusted by morphoclimatic weights (taking account of local topography and climate) or synoptic weather reports if available. BIAS has been applied to case-study situations in North and South America, Europe, Asia and Africa (Barrett, 1987; Barrett et al. 1985a, 1985b).

Interactive techniques for the estimation of 10-day or longer period estimates also include the FAO method and that of Callis and LeComte. In the FAO method, applied over Africa to assist the FAO's desert locust and drought-monitoring programme, low-resolution (spatial, radiometric and temporal) secondary data are analysed interactively to produce maps of 'Probably Precipitating Clouds' on a coarse spatial (1° latitude/longitude) grid; then 10-day estimates are produced by the multiplication of estimated number of days with probably precipitating clouds and the maps of mean rainfall per rainday (Barrett and Harrison 1986; Hielkema et al. 1987). The Callis and LeComte technique was also developed for application over Africa: more specifically, over the eastern Sahelian area. In this technique, an analyst interactively outlines the areas of thunderstorms and showers on 6-hourly Meteosat facsimile recorders. Regression equations are then used to make 10-day or longer period estimates of rainfall (Callis and LeComte 1987; LeComte et al. 1988).

TAMSAT at the University of Reading. For details of PERMIT the reader is directed to Chapter 5. The techniques can be divided into short period (daily to 5-day) and long period types (10-day to monthly).

Rainfall estimates are produced with no analyst interaction at all. Most of the automatic methods incorporate some form of thresholding (e.g. temperature thresholding, visible/infrared bi-spectral thresholding) or a form of automatic pattern recognition. They also include automatic measurements of various parameters, such as change in area with time, as the basis for their estimates, rather than human judgement. Lovejoy and Austin (1979a and b) combined geostationary satellite visible and thermal infrared data with a radar calibration set and found that clouds with the coldest tops produced the heaviest rain and defined the threshold requirement for precipitation as a cloud top colder than $-12\,^{\circ}C$ (261 K). This threshold produces a good general assessment of the distribution of precipitation over an area. Seven methods can be identified (Table 4.3).

1   Cloud-indexing methods: Cloud images are classified into cloud-cover areas of different meteorologically derived index values according to the probability of their producing rain and the nature of the rain produced. These methods extend the cloud-cover mapping procedures described in the last section.

2   Rainfall climatology methods: The relationship between climatologically averaged rainfall and the long-term average contribution made to it by numbers of key types of synoptic weather systems as identifiable from satellite imagery.

3   Life-history methods: These are based on two basic phenomena: that significant precipitation generally falls from convective clouds and that these clouds can be readily identified on satellite

Table 4.3 Summary of main groups of current satellite rainfall method types

Method	Chief applications	Satellite sensor(s)	Bands
Cloud indexing	Meteorology climatology Hydrology crop prediction, hazard monitoring	Polar-orbiting, geostationary	Visible and/or infrared
Climatological	Crop prediction	Polar-orbiting	Visible and/or infrared
Life history	Severe storm assessment meteorological research	Geostationary	Infrared
Bi-spectral	Meteorological research	Polar-orbiting, geostationary	Visible & infrared
Cloud physics	Cloud research & developmental atmospheric research	Geostationary	Infrared, microwave
Passive	Oceanic meteorology	Polar-orbiting	Microwave
Active	Cloud & rainfall research forecasting	Polar-orbiting	Satellite radar

Source: Modified from Barrett 1982 and 1987.

images. Rainfall can then be estimated by inference from past case-study assessments of similar cloud situations and life cycles.

4 Bi-spectral or tri-spectral methods: Objective spectral analyses of visible and infrared and/or Meteosat water vapour channel cloud images are used to map the extent and distribution of precipitation. These can then be related to conventional synoptic station data to give rainfall estimates.

5 Cloud model methods: These methods endeavour to develop 'human-machine' models which eluci-date the intricate relationships between cloud patterns and rainfall.

6 Passive microwave methods: Radiometer measurements by the Nimbus research satellites have produced some data which have been successfully processed to yield mesoscale rainfall-intensity distributions over sea areas but less successfully over land.

7 Active microwave methods: The launch of the TRMM satellite (Tropical Rainfall Measuring Mission) in 1997 has heralded the era of space-borne active radars specifically designed for rainfall determinations.

Automatic techniques can also be categorised according to their length of monitoring period and spectral wavebands used:

Short-Period Infrared-Based Techniques

A number of these techniques have been developed which use cloud top temperature measurements to make estimates of rainfall. There are difficulties in distinguishing, for example, between cold tops of deep clouds (such as cumulonimbus), which may precipitate heavily, and the equally cold tops of high, but relatively shallow clouds (such as cirrostratus), which may precipitate lightly or not at all. There is also a tendency to underestimate heavier rainfall amounts, for example Robertson's NASA/Marshall technique (1985). Some of the more sophisticated techniques shown in Table 4.2 consider additional parameters such as cloud area and rate of cloud growth with time.

Short-Period Bi-spectral Techniques

These techniques were developed to reduce the estimation errors encountered by the infrared-only techniques, by adding information during daylight hours. They use the basic assumption that if a cloud appears bright in the visible and cold in the thermal infrared then it is a raincloud. Inoue (1987) found that, while both cirrus and cumulonimbus clouds may be cold and bright, the spectral dependence of the emissivity of ice and water clouds differs within the thermal infrared window. By comparing the difference in response between the two bands (at 11 and 12 μm) on the AVHRR instrument he found that the two cloud types could be readily discriminated.

Tsonis and Isaac (1985) devised a method which could delineate instantaneous rain areas (in the mid-latitudes) from GOES visible and infrared images alone, without the need for coincident weather radar data. This method essentially employs a clustering analysis to locate peaks of response in the visible and infrared, which are attributed to raining and non-raining clouds.

Short-Period Pattern Recognition Techniques

These techniques employ pattern recognition algorithms applied either to a single image (infrared) or a pair of images (visible and infrared). Wu et al. (1985) developed an algorithm for tropical storms and cyclones near southern Florida and in the Gulf of Mexico coastal region. Their algorithm employed textural and radiance features (calculated over grid squares of 20 km $\times$ 20 km) in a hierarchical decision tree. For all case studies, the addition of texture features along with radiance features raised the accuracy in identifying rain, and therefore the classification of three rain rate classes.

Adler and Negri (1988) developed a method (known as the Convective-Stratiform Technique (CST)) which applies pattern recognition subroutines based on radiance and texture to infrared imagery, to identify convective and stratiform areas of clouds. A major disadvantage of this technique, however, is that it is very machine intensive.

Long-Period Automatic Techniques

Researchers at the other end of the time/space scales (i.e. long periods, large areas) have found that spatial and/or temporal averaging provides a useful means of simplifying the inherent complexities of the relationships between cloud properties and rainfall. Workers in the USA and Europe have spent the past few years applying long-term automatic techniques with the aim of improving our understanding of rainfall variation in different parts of the globe.

Arkin (1979) and Richards and Arkin (1981) found that for the GATE (GARP Atlantic Tropical Experiment) region (22.25–24.75°W longitude and 07–10°N latitude) there is a high correlation between rainfall and the fractional coverage of cold clouds (< 235 K) when both were averaged over areas of 1.5° × 1.5° or larger, but that this relationship did not hold for areas of 0.5° × 0.5° or smaller. They also found that while correlations were relatively high, even for hourly measurements, they increased for averages over longer time-periods. For the present technique, (Global Precipitation Index), rainfall in 2.5° latitude/longitude squares is estimated as a linear function of daily percentage coverage of each grid square by cold cloud. This method is only applicable to the tropics and the warm season in mid-latitude regions with suitable quantitative calibration to obtain estimates of rainfall.

Meisner and Arkin (1987) analysed spatial and annual variations in the diurnal cycle of large-scale tropical convective cloudiness and precipitation from the American GOES East and West satellites. They found that the diurnal pattern of convection in summer was strong, with a maximum of convection at 1,800 LST over the interior of South America and substantial parts of the United States; there was a much weaker cycle in winter, a nocturnal maximum being common. This precipitation index has also been incorporated into the range of routine products from the Meteosat geostationary satellite ESOC (European Space Operations Centre) Precipitation Index (Turpeinen et al. 1987). Verification results carried out with surface station reports from several African countries indicated that rainfall estimates (5-day averages) had a correlation of about 0.7 with observed values in tropical areas, but showed poorer correlations in the sub-tropics (Turpeinen et al. 1987). This better understanding has contributed to the improvement of operational weather forecasting.

An Example of a Long-Period Automatic Technique (PERMIT)

PERMIT is an example of a long-period infrared-based technique which was developed to estimate rainfall over the western Sahel. It uses four thermal infrared images a day from either GOES or polar-orbiting satellites. Daily rain-cloud areas are identified as those below a temperature threshold for rain-producing cloud and aggregated up for 10-day or longer periods. PERMIT has been further developed by Bellerby and Barrett (1993), using the Progressive Refinement Technique to enable reliable rainfall estimation for periods shorter than 10 days. The reader is directed to Bellerby and Barrett (1993) and Chapter 5 for more details of this technique and its application to drought monitoring in West Africa.

An Example of a Long-Period Bi-spectral Automatic Technique (ADMIT)

ADMIT is a long-period bi-spectral technique which uses visible and thermal infrared data to differentiate between dry and wet areas. Rain-clouds are identified as being bright (thick) in the visible and cold (high-altitude tops) in the thermal infrared. Daily 10-day and longer period maps of rain-cloud are produced and rainfall estimates derived by multiplying these rain-cloud areas by spatial maps of mean rainfall per rainday amounts, in a similar way to PERMIT.

This technique was developed for automated rainfall estimation over the whole of Africa using Meteosat data. It consists of the daily delineation of rain/no-rain areas by visible/infrared bi-spectral thresholding during daylight and temperature thresholding during night-time. Ten-day initial estimates are obtained by the multiplication of estimated number of raindays and morphoclimatic weight maps. Adjusted 10-day or daily estimates have been obtained when synoptic data have been available (Barrett and D'Souza 1986). Some tropical storms have lifetimes of less than 3–4 hours, and therefore more frequent images are required than those obtained by PERMIT/ADMIT.

An Example of an Automatic/Objective Technique (TAMSAT Cold Cloud Statistics Method)

This is another long-period thermal infrared technique. The TAMSAT method addresses the short-duration storm problem by using as many thermal infrared images as possible for each day to derive total duration of cold cloud below a certain threshold per day. To derive a Cold Cloud Duration (CCD) map, each thermal infrared image is scanned and all pixels with temperatures below the threshold are assigned a unit score. At the end of the required time period the units are added for each pixel (Bonifacio 1991). These are aggregated up for 10 days to give a good indication of duration of squall line activity over a given pixel or group of pixels. The temperature threshold used is high enough to include all precipitating cloud and low enough not to include warmer non-rain cloud. This technique has been adapted to run operationally and automatically with minimum human intervention by TAMSAT and the University of Bradford and is used in conjunction with PC-based meteorological satellite receivers in developing countries by the Natural Resources Institute.

The TAMSAT method has also been used operationally by the FAO within its ARTEMIS (Africa Real Time Environmental Monitoring Information System) Programme which supports the UN Food and Agriculture's drought early-warning function and desert locust control and survey activities.

This technique has been developed primarily to make 10-day, or longer period, estimates of rainfall in Africa. It relies on the correlation found between hours of cold cloud duration (as inferred from hourly Meteosat infrared imagery) and 10-day reported rainfall over Niger and other parts of the western Sahel. Very cold temperature thresholds have been adopted to identify cold clouds ($-50\ °C$ or $-60\ °C$) in case studies over various parts of the Sahelian region of Africa. It has been found that the relationships vary at different times of the year and in different regions (Milford and Dugdale 1987; Dugdale and Milford 1986; Flitcroft et al. 1987).

ARTEMIS (Africa Real Time Environmental Monitoring Information System)

In 1988 the UN FAO set up ARTEMIS to provide information for its famine early-warning services, namely GIEWS (Global Information and Early Warning System). It is an operational system providing environmental assessments in near real-time based on rainfall and vegetation condition monitoring using low spatial resolution meteorological satellite data. Initially, ARTEMIS used Meteosat data to produce cold cloud duration and estimated rainfall images and NOAA AVHRR normalised difference vegetation index images over Africa for 10-day and monthly period estimation. An ARTEMIS product for two consecutive years for central Africa is shown in Plate 4.1. By the end of 1998, images covering South/Central America and Asia were being used as well as those from Africa at resolutions of 1–8 km. ARTEMIS has produced a series of analysis and image display tools that are PC and workstation compatible in conjunction with its data products (Snijders and Minamiguchi 1998).

4.5 WEATHER FORECASTING SYSTEMS

Weather forecasting systems are based on interactive precipitation techniques. Fully operational continuous weather forecasting systems are operated by the NOAA Environmental Satellite Data and Information Service (NESDIS) and the UK Meteorological Office with their Nimrod forecasting system. The interactive precipitation estimation technique originates from the Scofield–Oliver technique (Scofield and Oliver 1977), which was initially devised as a subjective manual technique based on hardcopy visible and infrared imagery for the estimation of rainfall from deep convective systems. Sequences of half-hourly images were examined and rainfall estimates computed as a function of cloud top temperature and the presence or absence of overshooting tops and mergers. The technique has undergone many improvements and refinements, and many aspects of it have been automated such as the Interactive Flash Flood Analyzer (IFFA) (Clark and Borneman 1984; Clark and Perkins 1985). This is used operationally for real-time precipitation estimation of heavy rainfall over the United States of America. This technique has also made an important contribution to the location of heavy snow and rainfall during the winter season, which is essential for issuing flash flood warnings (Scofield 1986).

Lovejoy and Austin (1979a) developed a technique whereby real-time radar data are used to derive 'optimum threshold curves' which delineate raining areas on simultaneous visible (0.4–1.1 μm) and thermal infrared (10.5–12.5 μm) images by two-dimensional thresholding. This procedure has been found to work well in separating low, thick or high, thin non-precipitating clouds from cumulus systems typical of many mid-latitude synoptic situations. They also found that GOES thermal infrared and visible imagery could be used for the estimation of both rain area and amount estimation over long time periods and large areas, but was poor in rain rate determination. The technique has been used quasi-operationally in two real-time (so-called 'nowcasting') rainfall estimation schemes,

the large-scale RAINSAT in Canada (Bellon *et al.* 1980 and Austin and Bellon 1982) and the FRONTIERS system (Forecasting Rain Optimised using New Techniques of Interactively Enhanced Radar and Satellite) in the United Kingdom (Browning 1979; Browning and Collier 1982). This enables the forecaster to blend data from several radars with Meteosat imagery and rain-gauge data to produce effective detailed half-hourly initial real-time nowcasts for a few hours ahead. FRONTIERS was the precursor to the Nimrod nowcasting system, currently used by the UK Meteorological Office.

Nimrod

This is the current mesoscale nowcasting system used for preparing the UK Meteorological Office weather forecasts for the British Isles and northwest Europe. It has been operational since 1995. It is a fully automatic system combining ground radar, weather satellite and other observational data to produce analyses and forecasts of rainfall and other weather parameters for up to six hours in advance. Radar data provide rainfall updates every 15 minutes at a 5 km spatial resolution (in some areas where there are better radar instruments, 2 km resolution data is given at 5-minute intervals). These ground radar measurements are regularly compared with data from hourly recording rain-gauges.

Meteorological satellite images are used to extend the rainfall analysis beyond the range of the ground radar. Infrared and visible images from Meteosat are half-hourly and are used to produce rainfall area and amount estimates for this purpose. Nimrod products include precipitation forecasts of amount and type, low cloud and visibility reports (UK Meteorological Office 1997).

4.6 DISCUSSION

Clearly, a wide variety of techniques for cloud analysis and rainfall estimation from satellite visible and infrared data have been developed for a wide variety of applications and to suit a large number of end-users. The various principles underlying satellite-based rainfall estimation techniques have depended not only on the application to hand but also on the meteorology of the area of application, the type and availability of satellite and/or ancillary data, the type of processing equipment or manpower and on the type of verification data available. In general, techniques have been developed which:

- may analyse satellite imagery for cloud type, cloud distribution and cloud systems and/or for rain-cloud and rainfall estimation;
- may be manual, interactive or automatic;
- may produce short-period estimates throughout the stage of development of convective systems, or they may produce daily or 10-day or even longer period total estimates of rainfall;
- may vary in the spatial size of the estimates from 1 to 2 km to 2.5° latitude/longitude squares (maximum of c. 290 km at the equator);
- may be based on mono-spectral or bi-spectral data;
- may be based on half-hourly to once daily satellite imaging frequency;
- may be based entirely on satellite data or may incorporate a number of other remotely sensed or *in situ* data;
- may be verified against hourly point or area estimates of rain rate or amount, or daily, 10-day, or monthly point rainfall totals.

This makes comparative evaluations of each technique difficult, not least because many have been applied in very different locations and conditions, and because most have undergone only limited (and not directly comparable) verification. However, the verification study of Snijders (1991) has highlighted some important questions to be answered in satellite-based rainfall estimation. It would seem that a method which incorporated visible as well as infrared imagery would delineate rain areas more accurately than the methods which used infrared imagery alone. Similar conclusions have been drawn by Wu *et al.* (1985), Bellon and Austin (1986) and Tsonis (1987), among others. Snijders (1991) and Negri and Adler (1987a and b) also agree with this conclusion but argue that the improvement is relatively modest. However, it is notable that researchers and designers of the Tropical Rain Measuring Mission (TRMM) have decided the additional information is important. It has also been shown that, although infrared-only techniques are not as accurate as bi-spectral ones, they do significantly reduce the amount of data required and thus not only the cost of the operation but also the amount of computer-processing time without greatly affecting estimate accuracy.

The question of temporal sampling through the day has also received much attention. In his study on the effect of temporal sampling on the observation of mean estimated rainfall, Laughlin (1981) found that, for GATE, only four observations per day were sufficient to determine the mean monthly rainfall over an area of 2.5° square to within a standard deviation of 5 per cent of the mean value. Wylie and Laitsch (1983) studied the impacts of varying amounts of satellite data on rainfall estimation schemes using full-resolution GOES imagery over the Great Plains States (USA). They tested seven methods (which ranged from manual analysis of only two images per day without any other data, to using hourly imagery combined with ancillary data) and found that all produced very similar patterns of monthly accumulations. The daily rainfall patterns, too, reflected similarities among the seven methods tested. Snijders (1991) found that ADMIT (which used, on average, 4-hourly sampling) scored a higher Critical Success Index than methods which used hourly sampling.

Most techniques underestimate large falls of rain and overestimate small ones. In order to produce the most accurate estimate, it is necessary to incorporate any ground-station measurements and/or radar where and when these are available. However, distance-weighted interpolation techniques have been shown to be inaccurate at large distances from reporting stations. In this connection, the studies that have used the geostatistical interpolation techniques of kriging and co-kriging seem to be more promising (see, for example, Lebel 1986; Creutin and Delrieu 1986; and Krajewski 1987). The GARP Atlantic Tropical Experiment (GATE) radar precipitation analysis provides a high-quality dataset resolving the structure of convection over the eastern tropical Atlantic Ocean and thus ideal ground verification data for this study. Research results, in establishing an empirical relationship between thermal infrared cloud cover and rainfall, suggest that for reasonably large timescales and space scales, such simple methods may have sufficient accuracy for large-scale climatic studies.

In terms of the new design and implementation of operational satellite-based visible/infrared rainfall estimation techniques over large regions, it is clear that no single technique would be able to provide rainfall estimates at all the spatial and temporal scales required. It seems likely that, in this case too, there will be some integration of available techniques. Over the past few years several national meteorological services have endeavoured to produce integrated forecasting systems incorporating different temporal and spatial scales of weather information. The design and development of microprocessor systems which incorporate all the various hierarchical and integrated automatic and interactive routines is already well under way.

PART II: VEGETATION MONITORING FROM SATELLITE

4.7 INTRODUCTION

Much of the Earth's land surface has an uppermost cover of vegetation which intercepts electromagnetic radiation transmitted through the atmosphere. As a consequence, satellite images obtained over land surfaces invariably contain areas of vegetation cover. Remote sensing of vegetation is important in the following areas:

1 Natural vegetation resource inventory
2 Land-cover assessment – mapping and monitoring particularly in remote areas
3 Monitoring planted crops
4 Detection of diseased and insect-infested crops.

The electromagnetic response from vegetated surfaces will be considered first, followed by a discussion of the methods by which remotely sensed images can be enhanced for the extraction of vegetation information. Finally, the application and interpretation of these enhancements are addressed. Classification of land-cover types was discussed in Chapter 3. This section of Chapter 4 will focus on how data can be processed and enhanced for vegetation information extraction and interpretation. However, many of the enhancements can be used to create additional bands which can improve the accuracy and performance of image classification in vegetated areas.

4.8 VEGETATION RESPONSE

Leaf Structure and Spectral Response

An understanding of the spectral properties of individual leaves and plants is vital for the interpretation of images of vegetated landscapes. The cuticle and epidermis form the outermost cell layer of a leaf and are composed of specialised translucent cells which can be penetrated by electromagnetic radiation (Figure 4.4). The lower epidermis has pores called stomata which allow carbon dioxide to enter the leaf for photosynthesis but they also maintain the thermal balance and respiration of the leaf by permitting moisture and gaseous exchange. The palisade cells are beneath the upper epidermis and contain chloroplasts with chlorophyll, which is a green pigment that absorbs the red and blue elements of white light but reflects the green element. Therefore healthy vegetation appears green in the visible range and it is this layer of cells that is responsible for most of the response from the leaf in the visible part of the spectrum. A leaf reflects about 10–30 per cent of total light falling on it in the green part of the spectrum.

The next layer consists of spongy mesophyll cells,

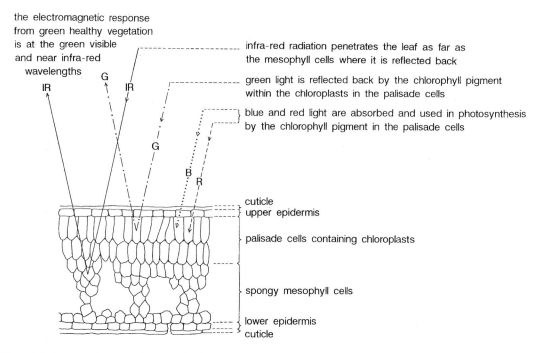

the electromagnetic response
from green healthy vegetation
is at the green visible
and near infra-red
wavelengths

IR   G   IR

G

B   R

infra-red radiation penetrates the leaf as far as
the mesophyll cells where it is reflected back

green light is reflected back by the chlorophyll pigment
within the chloroplasts in the palisade cells

blue and red light are absorbed and used in photosynthesis
by the chlorophyll pigment in the palisade cells

cuticle
upper epidermis

palisade cells containing chloroplasts

spongy mesophyll cells

lower epidermis
cuticle

Figure 4.4 Cross-section through a green leaf showing spectral responses from its different cell layers. Modified from Campbell (1996).

which are irregularly shaped cells with a large surface area. These cells reflect about 60 per cent of the near-infrared irradiation back from the leaf. The internal structure of leaves plays a key role in the bright near-infrared response of living vegetation, the peak reflectance of vegetation being in the near-infrared not green part of the electromagnetic spectrum. Therefore the near-infrared bands of satellite and air-borne imagery are very useful in vegetation discrimination. The response in the near infrared enables the separation of vegetation from non-vegetated surfaces, which are usually darker in the near infrared. Differences in reflectivities of plant species are more pronounced in the near infrared than in the visible green; the discrimination of different vegetation types is thus often better in the near infrared than in the visible. However, it is useful to take note of the Earth surface response in the visible red too, as a high response in both the visible red and near infrared indicates that the surface is not vegetated but likely to be an artificial anthropogenic surface.

As a plant senesces or becomes stressed or diseased, cells die, change in structure and lose their chlorophyll pigment. The spectral characteristics of the leaf therefore change in both the visible and near infrared. In the near infrared the reflection is reduced by cell deterioration and in the visible it increases. This is shown by the spectral response curves in Figure 4.5. Thus near-infrared bands are useful for monitoring vegetation vigour and detecting and mapping the areal extent of vegetation deterioration that is due to disease/insect infestation. The near infrared can also be used to detect natural senescence associated with ripening crops. At longer infrared wavelengths (greater than 1.3 μm), water content controls the spectral response of a leaf. Longer wavelength radiation, between 1.4 and 2.5 μm, is absorbed by the leaf.

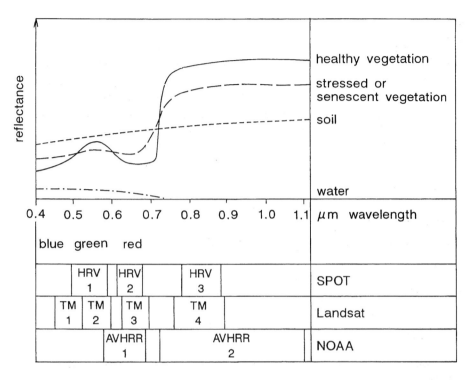

Figure 4.5 Response curves of healthy and senescent vegetation from the visible to long infrared, compared with response curves from vegetation canopies and soil surfaces. Lower part of figure shows location of TM, SPOT (1, 2 and 3) and NOAA bands.

Vegetation Canopy Response

This is more complex than the response from a single leaf and is more likely to be encountered in reality. Canopies are composed of many layers of many leaves at different heights above the ground, and are also of different sizes and orientations. Leaves higher in the canopy form shadows over leaves below, so that overall reflectance is a combination of leaf reflection, re-reflection off upper and lower layers, and shadow, which depends on illumination angle and leaf orientation.

Shadows and within-canopy re-reflection reduce the returning top-of-canopy reflection by as much as 50 per cent of the single-leaf reflection in the visible, and by 70 per cent in the near infrared. The relative decrease in the near infrared is less than in the visible, because near-infrared reflection is transmitted back through the canopy to the upper leaves so that it adds to the returning electromagnetic reflection which is picked up by the satellite sensor whereas the visible wavelengths are unable to penetrate right through from the underside of the leaf. Some of the reflection in the near infrared may be from soil as well as leaves (see Figure 4.5).

Detection of Mature Vegetation and Senescence

When plant cells begin to alter as the plant starts to ripen or senesce, there are changes in its spectral response. The first difference is a general increase in the visible spectral response as cells containing chloroplasts disintegrate and the chlorophyll pigment becomes inactive and therefore unable to absorb the blue and red components of white light. Second, there is the 'red shift' first noticed by Collins in 1978, when he carried out a spectral response survey using Landsat MSS data of many crops at different stages of growth. He concentrated on their responses at the far visible red part of spectrum, which is where chlorophyll absorption decreases and infrared reflection increases. In this region of the spectrum from just below 0.7 μm to just above 0.7 μm, vegetation response increases by about an order of magnitude (Collins 1978). As the plants approach maturity the position of the chlorophyll absorption edge on the spectral response curve shifts towards the longer wavelengths. This is known as the 'red shift' (Campbell 1996). The causes of the red shift and its variation between plant species are very complex and are still under research, but it is one way in which senescent or stressed crops or vegetation may be detected, often earlier in their growth cycle than by visual means alone.

Separation of Soil from Vegetation Response

The reflectance measured by a sensor often has a soil and vegetation component. With natural vegetation there is less mixing, but in semi-arid and arable crop areas the response is always mixed in the early stages of growth. Information on vegetation can only be extracted when coverage is at least 30 per cent.

If the red and near-infrared responses of many different bare soil surfaces are plotted on a scatter diagram, soil response will form a more or less straight line because, as response becomes brighter in the visible red, it also becomes brighter in the near infrared (Figure 4.6). This is the 'soil brightness line', which was first devised by Richardson and Wiegand (1977). Wet soils are at the bottom of the line as they tend to be darker than dry soils, which are found at the high end of line. However, soil response also depends upon the mineral and organic material components of the soil. For instance, a soil over chalk bedrock will have a much brighter response (Figure 4.6) in the visible and the near infrared, whereas a highly organic soil such as peat will have much lower responses in both

these bands (Figure 4.6). Reflectance in the near infrared is more sensitive to changes in green vegetation on dark substrates than on light ones, whereas reflectance in red is more sensitive to changes in the green vegetation on light-toned substrates. (The reader can create a soil brightness line with the data and image processing software on the CD accompanying this text.)

Spectral response of living vegetation always has a consistent relationship to the line. This line is one of the basic assumptions on which vegetation indices are developed and in many studies the soil line is considered to be the zero vegetation line. Soils have a fairly high response in visible red and infrared wavebands whereas healthy vegetation has low values in the visible red and high values in the near infrared. A healthy vegetation stand is thus represented by A in the upper left of Figure 4.6. Points B1 and B2 represent partly vegetated pixels with different proportions of vegetation and soil.

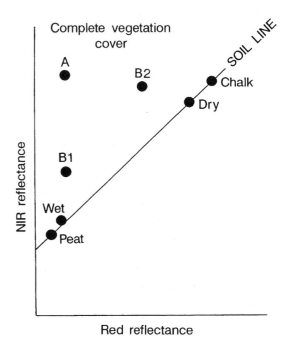

Figure 4.6 Soil Brightness Line. After Richardson and Wiegand 1977; modified from Campbell (1996).

4.9 IMAGE ENHANCEMENTS TO EMPHASISE VEGETATION

Band Ratios

Ratio images were introduced in section 3.4 and the reader should be familiar with this material before continuing. Ratio images give a measure of the difference in reflectance of the same surface for two separate portions (bands) of the electromagnetic spectrum. They are often used to expose hidden information, particularly when there is an inverse relationship between two spectral responses to the same Earth surface. They are therefore used in geological mineral differentiation as well as for discriminating vegetated from non-vegetated surfaces. If two different surfaces have the same spectral response in two electromagnetic bands, then ratios provide little additional information to that gained from examining the surface by using one band alone. Ratioing is particularly useful in vegetation detection because of the high spectral absorption in the visible red and high reflectance in the near-infrared region. The infrared/red ratio produces a high pixel value when vegetation is present but a low value for any other surface when vegetation is not present. Figures 4.7a and b are TM 3 and TM 4 images respectively for the Tavistock region of England. The healthy cultivated vegetation in the west is much brighter than the poor moorland vegetation in the east on the TM 4/TM 3 ratio image (Figure 4.7c). The town of Tavistock has a very low ratio value. This is the simplest form of vegetation index and is often referred to as the RVI (Ratio Vegetation Index; Jordan 1969). This index gives a value between 0 (no vegetation) and infinity (complete healthy vegetation cover) which gives a measure of the importance of vegetation reflectance in a pixel. Sometimes the green visible/red visible ratio is used, but it is not as effective because the difference between the response of vegetation in the green visible band is not as great as in the infrared.

Vegetation Indices

Vegetation indices are empirical formulae designed to emphasise the spectral contrast between the red and near-infrared regions of the electromagnetic spectrum, i.e. around 0.7 μm. They produce digital quantitative measures, which attempt to measure biomass and vegetative health (Campbell 1996): the higher the vegetation index value, the higher the probability that the corresponding area on the ground has a dense coverage of healthy green vegetation. They can be regarded as a more sophisticated form of the ratio of the near-infrared and red bands of the spectrum. A vegetation index can also be regarded as a mechanism by which image band data can be combined or compressed (by reducing the relevant red and near-infrared band information into a single-band image) so that future processing, for instance classification, becomes more efficient.

Several different vegetation indices have been developed, one of which has already been introduced, the RVI. This is the simplest index and is a ratio between two digital bands. Much initial vegetation index research was carried out using Landsat MSS imagery (Campbell 1996). However, Landsat Thematic Mapper bands 3 and 4, NOAA AVHRR channels 1 and 2 and SPOT bands 2 and 3 are now more commonly used in vegetation studies.

In all vegetation indices two basic assumptions are made: first that an algebraic combination of remotely sensed spectral bands can provide useful information about vegetation, and second that all bare soil in an image will form a line in spectral space, i.e. the soil brightness line as the line of zero vegetation. After this, there are currently two divergent schools of thought about the orientation of the lines of equal vegetation cover (isovegetation) in relation to the soil brightness line: first, that all isovegetation lines converge at a single point. The ratios employing this assumption are the ratio-based indices, which measure the slope of the line between the point of convergence and the red/near-infrared point of the pixel, for example the Normalised Difference Vegetation Index, Soil Adjusted Vegetation Index and Ratio Vegetation

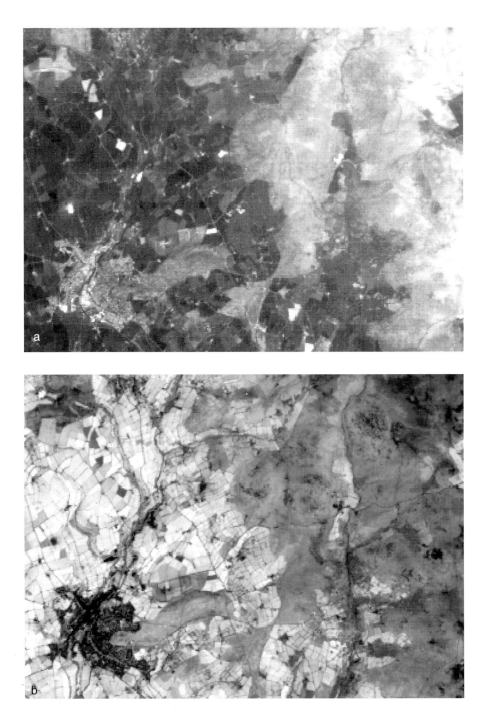

Figure 4.7 (a) TM 3 image of Tavistock area, England. (b) TM 4 image of Tavistock area England. Original data from European Space Agency 1988, distributed by Eurimage/NRSC.

Figure 4.7 (c) TM 4/TM 3 ratio. Original data from European Space Agency in 1988 and distributed by Eurimage/NRSC.

Index. The second school of thought is that all isovegetation lines remain parallel to the soil line. These indices measure the perpendicular distance from the soil brightness line to the red/near-infrared point of the pixel, for example the Perpendicular Vegetation Index, Weighted Difference Vegetation Index and Difference Vegetation Index (Ray 1996).

Perpendicular Vegetation Index (PVI)

Richardson and Wiegand (1977) quantified the difference between soil and vegetation reflectance by defining the Perpendicular Vegetation Index, PVI, which is a measure of the distance of a pixel from the soil brightness line (Campbell 1996).

$$PVI = \frac{1}{\sqrt{(S_r - V_r)^2 - (S_{ir} - V_{ir})^2}}$$

where S = soil reflection, V = vegetation reflection, r is red wavelength and ir is infrared wavelength. This produces an index value between −1 and +1. Ray (1996) gives an alternative formula for the PVI:

$$\sin (a) \text{ near infrared } 2 \cos (a) \text{ red}$$

where a is the soil line gradient.

Normalised Difference Vegetation Index (NDVI)

One of the most widely used vegetation indices is the Normalised Difference Vegetation Index (NDVI), which was first proposed by Kriegler *et al.* (1969).

$$NDVI = \frac{\text{near infrared } - \text{ visible red}}{\text{visible red } + \text{ near infrared}}$$

This, in principle, is similar to the near infrared/red ratio but is tailored to give desirable statistical characteristics for various parameters associated with vegetation growth, type and ecosystem environment. It produces an index value ranging between −1 (no vegetation) and +1 (complete healthy green vegetation cover). In this index the isovegetation lines converge at the origin of the bi-spectral near infrared/red scatterplot.

Difference Vegetation Index (DVI)

DVI = (near infrared − visible red)

This is an index with isovegetation lines parallel to the soil line and is often referred to as the VI, for example in Lillesand and Kiefer (1994).

Weighted Difference Vegetation Index (WDVI) (Clevers 1988)

WDVI = near infrared − g × red

where g is the gradient of the soil brightness line.

The WDVI is a mathematically simpler version of the PVI but has an unrestricted value range (Ray 1996).

In all the indices described above, the soil line is assumed to be a single line in the red/near-infrared feature space of a single image. However, even in the same image, where there are different soil types and colours, different soils have different red/near-infrared line slopes. This problem becomes more critical when vegetation cover is low. A series of indices has been derived to reduce this problem. These are ratio-based indices and also tend to be more sensitive to atmospheric variations.

Soil Adjusted Vegetation Index (SAVI)

This was developed by Huete in 1988 and is a hybrid between a ratio-based and a perpendicular index. This index allows for the isovegetation lines not being parallel and not all converging at a single

point. The adjustment factor (L) was found by trial and error to give equal vegetation index results for both dark and light soils. The isovegetation lines converge in the negative red and negative near-infrared quadrant of the near-infrared/red feature space plot. Index values vary between −1 and +1 (Ray 1996).

$$\text{SAVI} = \frac{(\text{near infrared} - \text{red})}{(\text{near infrared} + \text{red} + L) * (1 + L)}$$

where L is a correction factor of 0 for very high vegetation cover, 1 for very low cover and 0.5 for intermediate cover (* is multiply).

Transformed Soil Adjusted Vegetation Index (TSAVI) (Baret *et al.* 1989; Baret and Guyot 1991)

This is a ratio-based index with the isovegetation lines converging in the negative red and negative infrared quadrant of the feature space plot. The index ranges between −1 and +1, and assumes that the soil line has an arbitrary gradient and intercept, and that these values are used to adjust the index.

$$\text{TSAVI} = \frac{a\,(\text{near infrared} - a * \text{red} - b)}{b * \text{near infrared} + \text{red} - b * a + X * (1 + a^2)}$$

where b is the soil line intercept, a is the soil line gradient and X is an adjustment factor which is used to minimise soil noise (0.08), * is multiply (Ray 1996).

Modified Soil Adjusted Vegetation Index (MSAVI)

This index was developed by Qi *et al.* in 1994. It is a ratio-based index and the isovegetation lines cross the soil line at different points. The soil line has an arbitrary gradient and a range between −1 and +1 (Ray 1996).

$$\text{MSAVI} = \frac{(\text{near infrared} - \text{red})}{(\text{near infrared} + \text{red} + J) * (1 + J)}$$

where J is a function of s (the gradient of the soil line) and NDVI and WDVI.

$$J = 2 * S * NDVI * WDVI$$

Global Environment Monitoring Index (GEMI)

This was developed by Pinty and Verstraete in 1991 to allow for the varying effects of the atmosphere on the values of vegetation indices, particularly when multitemporal studies to compare seasonal vegetation growth are required. The index varies approximately between 0 and +1 over continental areas, with values of greater than 0.35 indicating vegetated areas, small positive and negative values for soils and strongly negative values for clouds (Flasse and Verstraete 1993).

$$GEMI = \eta * (1 - 0.25 * \eta) - \frac{red - 0.125}{1 - red}$$

$$\eta = \frac{2(near\ infrared^2 - red^2) + 1.5 * near\ infrared + 0.5 * red}{near\ infrared + red + 0.5}$$

4.10 APPLICATION AND INTERPRETATION OF VEGETATION INDICES

The intensity of vegetation reflectance is dependent on the wavelength used and the three components of the vegetation canopy, i.e. leaves, substrate/soil and shadow. The reflectance of soil/substrate depends on its nature, whether or not it is composed of a light-coloured mineral soil or dark-coloured organic material, presence of senescent or understorey vegetation, and shadowing caused by vegetation canopy and other features such as topography. All these factors need to be considered in the selection of appropriate vegetation indices and their application and interpretation.

Vegetation indices can be used to measure or detect various characteristics of vegetation cover. These include:

- spatial distribution analysis
- event detection e.g. drought, post-rain vegetation flush
- Leaf Area Index
- fractional vegetation cover
- absorbed photosynthetically active radiation and crop productivity
- stocks and fluxes of carbon
- detection of recent rainfall events.

There are three current major areas of research on vegetation indices: first, validation of their usefulness for determining biological properties of plants, second, on their role as mapping tools to define areas of vegetation from other Earth surfaces, and finally the refinement of their equations so that they can be adapted for use in a number of different global environments. The various forms of index discussed in the previous section are the manifestations of this latter type of work.

1 Validation of their Usefulness and Testing of how Closely they Resemble Different Biological Properties of Plants

This research aims to establish the use of vegetation indices for remote monitoring of the growth, productivity and seasonal fluctuations of crops and pasturelands. Leaf Area Index, which is the area of leaf surface per unit of soil surface, and biomass (the weight of vegetation tissue) assessment are the two biological parameters that have been most successfully determined from remotely sensed imagery to date (Campbell 1996).

Leaf Area Index (LAI)

Vegetation indices have often been compared with *in situ* measurements of LAI and can give important information on the amount of leaf area exposed to the atmosphere for photosynthesis and respiration. These are key factors in determining plant growth and health (Tucker 1979; Campbell 1996). LAI is negatively related to the red reflectance but positively

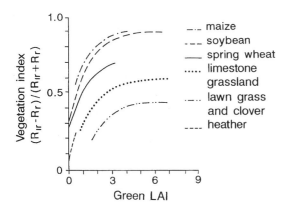

Figure 4.8 Relationship of Leaf Area Index to the NDVI (after Curran 1983)

related to near-infrared reflectance; a ratio of near infrared to red (i.e. the vegetation index) therefore expresses the increasing difference between red and near-infrared reflectance with increasing green LAI. The relationship between LAI and VI is curvilinear, as shown in Figure 4.8.

Biomass

Vegetation indices have been quantitatively compared to biomass measurements in a number of research studies. Several vegetation indices closely reflect the biomass of specific crops but no single index is effective for all crops/plants in all atmospheric conditions. Vegetation indices are useful for quantitative assessment of biomass in proportional terms but not in defining specific amounts and relationships.

2 Vegetation Indices as Mapping Tools

Vegetation indices are often used as qualitative tools for mapping vegetation, although sun and sensor look angles should be considered in this application because of shadow effects. Vegetation indices are also used qualitatively to assist image classification, to separate vegetated from non-vegetated areas and to assist in more accurately defining different types and

densities of vegetation (Campbell 1996). They are often generated over the same area at various time intervals to monitor seasonal or year-on-year variations in vegetation health, abundance and distribution (see Chapter 5). An example of this monitoring application is a NOAA AVHRR Global Vegetation Index (GVI) product. This is a Normalised Difference Vegetation Index applied to the low spatial resolution imagery of the NOAA AVHRR sensor. Although the spectral channels 1 (red) and 2 (near infrared) of AVHRR have slightly different wavelength ranges from those of Landsat MSS and TM, the meaning of the resultant NDVI is still the same. Resolution is coarser but coverage is much broader and repeat coverage is greater and therefore less likely to suffer from cloud cover.

The GVI provides vegetation information on a continental or subcontinental scale to give a broad geographical perspective of vegetation and enables observations of major ecological zones and seasonal changes, which were not possible before satellite technology (see Plate 4.2). This plate shows a coloured GVI product of northwest Europe which highlights the desert and semi-arid regions of northern Africa and southern Spain in browns and yellows, grading through pale and dark green northwards to Scandinavia and the British Isles.

Since 1982, NOAA has supplied weekly GVI products, and other organisations such as the Joint Research Centre of the European Union in Italy and UN FAO also supply periodic composite NDVI products from AVHRR to monitor regional and continental vegetation status world-wide. These are now being used in major global environmental change research programmes such as the International Geosphere Biosphere Project (IGBP). The resolution is too coarse to show individual crops/fields but these products do give a good view of overall crop cycles, seasonal patterns of irrigation in semi-arid areas, crop maturity and harvest. They can be combined to show vegetation dynamics and are excellent for providing estimates of major regional contrasts in vegetation activity but they should not be used to evaluate local trends (Goward *et al.* 1993).

3 The Refinement of Existing Vegetation Indices to Improve the Quantitative Measures that can be Derived from them

Tucker *et al.* (1985) at NASA Goddard Space Flight Center in Maryland, USA, have established that there is little difference in practice between the vegetation indices that have been developed. However, they should be used with care as there are many factors external to plant leaf structure and growth that can alter the ratio value, e.g. soil/row spacing in crops, atmospheric conditions, low vegetation cover. The type of study required is important in the choice of the most suitable index, for example whether the imagery is high or low spectral or spatial resolution data and whether or not the study is to be multitemporal.

Atmospheric Conditions

Vegetation indices can be sensitive to the atmospheric degradation of the visible bands, which are more vulnerable to atmospheric attenuation than the near-infrared ones and this therefore alters the value of the ratio. Vegetation indices should be compared over the same area over a period of time. The NDVI and WDVI are very sensitive to atmospheric variations whereas other indices are less so.

Low Vegetation Cover

The soil-adjusted variations of the vegetation indices are best used when plant cover is low, e.g. SAVI, MSAVI and TSAVI are all effective for vegetation assessment above 15 per cent, whereas the NDVI, RVI and DVI are only really useful when vegetation cover is at 30 per cent or more. The PVI and WDVI are also effective down to levels of 15 per cent (Ray 1996).

High Spectral Resolution Data

Specialised vegetation indices have been developed for higher spectral resolution data provided by airborne sensors. The reader is directed to Elvidge and Chen (1995). However, for satellite applications, SAVI or the PVI are considered better than the RVI and NDVI.

For broad-band sensors such as the Landsat Thematic Mapper and Multispectral Scanner the NDVI is still the most effective general-purpose index for all conditions, particularly for qualitative study, then the PVI, followed by SAVI (with a 0.5 correction factor when vegetation cover is low) and then MSAVI. The drawback with the indices that correct for soil background is that they can be inaccurate if no atmospheric correction has been applied. For multitemporal studies, atmospheric corrections should always be applied before the chosen vegetation index is used, and the variability of the soils of the area should be taken into account when the choice of index is made (Ray 1996).

Flasse and Verstraete (1993), using AVHRR data over Africa, found that GEMI is less sensitive to atmospheric conditions than the NDVI but is much better at detecting vegetation when it is sparse. It is also good for detecting and characterising clouds, and is highly sensitive to medium and bright soils.

4.11 FEATURE EXTRACTION FOR CROP GROWTH AND AGRICULTURAL APPLICATIONS

Phenology is the relationship between vegetation growth and environment, i.e. seasonal changes in vegetation growth and decline. Plants change shape, structure and appearance during their growth cycles and particularly in senescence, which is the progressive deterioration of leaves, stems, fruit and flowers. Cell walls deteriorate in the mesophyll tissue, which produces a sharp decline in infrared reflection (Campbell 1996). The reduction of chlorophyll as an absorber of visible red and blue radiation causes the red shift; senescence is therefore readily detected by Earth observation satellite sensors.

Remotely sensed data can be used to provide overviews of vegetation communities, for example the geographical spread of new growth in spring is a 'green wave' and late summer senescence is a 'brown wave' just before harvest. Regular imagery throughout a growing season will show 'green and brown wave' movement across continents. NOAA AVHRR is better for these general studies than Landsat because it produces imagery more frequently over a broader area.

In the agricultural context, the visual representation of phenology shown on satellite imagery illustrates the local crop calendar, which is the seasonal cycle of ploughing, planting, emergence, growth, maturity, harvest and fallow. Each region has a unique crop calendar determined by local climate and farming practices (Campbell 1996). Table 4.4 shows an example of such a crop calendar for the UK. The effects of crop husbandry activities can be seen on satellite imagery. Crop identification, classification and interpretation are often improved by a knowledge of the growth cycles of individual crops, and the ability to compare winter with spring crops. For example, in the British Isles winter wheat has bare fields in September and October, and mixed soil and newly emerged vegetation in late autumn or winter. In spring the crop growth increases to complete cover in early June, and senescence occurs in late June or July before harvest. A spring crop does not have complete cover until late June/July and is not senescent until August and September. Therefore by using several images obtained at different times throughout the season different crops can be discriminated. There are several examples of agricultural crop-monitoring projects using satellite imagery and the crop calendar, for example the US LACIE and AgRISTARS and the European CEC Agricultural Monitoring Project.

4.12 TASSELED CAP TRANSFORMATION

In 1976, Kauth and Thomas developed a linear transformation that projects soil and vegetation information into a single plane in multispectral data space, i.e. a two-dimensional graph in which the major spectral components of an agricultural scene are displayed (Campbell 1996). It was originally developed on Landsat MSS data but now has been further developed to accommodate all six non-thermal bands of Thematic Mapper data. It is often regarded as being a 'vegetation' equivalent of principal components analysis (see section 3.7).

The transformation can be visualised as the rotation of a solid multidimensional figure representing all spectral bands which enables an analyst to view the major spectral components of an agricultural scene as a two-dimensional figure, (Campbell 1996). When applied to Landsat MSS imagery, the transformation consists of linear combinations of the green, red and two near-infrared bands which produce a new set of bands: TC1, TC2, TC3 and TC4. The following equations have been adapted from Campbell (1996).

TC1 = + 0.433 (green) + 0.632 (red) + 0.586 (near infrared) + 0.264 (near infrared). This gives a measure of 'greenness', i.e. abundance and health of vegetation and is equivalent to PVI.

TC2 = − 0.290 (green) − 0.562 (red) + 0.600 (near infrared) + 0.491 (near infrared). This gives a measure of 'soil brightness' (Richardson and Wiegand 1977).

TC3 = − 0.829 (green) + 0.522 (red) − 0.039 (near infrared) + 0.194 (near infrared). This gives a measure of 'yellowness' or senescence.

TC4 = + 0.223 (green) + 0.012 (red) + 0.543 (near infrared) + 0.810 (near infrared). This is referred to as 'nonsuch' and is considered to contain system noise and atmospheric information, i.e. information that is of little use to a vegetation/crop assessment study.

As with PC1 and PC2 in principal components analysis, TC1 and TC2 carry about 95 per cent of vegetation information. Therefore, a two-dimensional graph of TC1 against TC2, as shown in Figure 4.9, provides most of the information required for agricultural image interpretation.

Figure 4.9 is a graphical representation of the Tasseled Cap transformation and demonstrates that,

Table 4.4 Crop calendar for the UK

Crop	Jan	Feb	Mar	Apr	May	Jun	Jul	Aug	Sep	Oct	Nov	Dec
Winter wheat												
Winter barley												
Winter oats												
Spring cereals												
Spring oilseed												
Winter oilseed												
Sugar beet												
Kale												
Grass (hay)												
Grass (silage)												
COVER		25–75%		100%		In flower		Ripe		Harvest		Bare soil

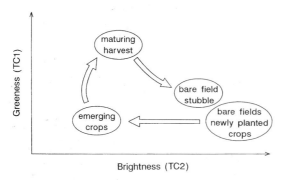

Figure 4.9 Variation in brightness and greenness response of a crop during a typical growth season cycle. Adapted from Crist (1983).

as a crop emerges and grows, soil response becomes less and green response more (progression clockwise from the bottom right to bottom left corner of the graph). As it matures, greenness gets brighter (top left of the graph), then senescence causes reduction in greenness and harvest finally increases in soil brightness again (bottom right of the graph). A graph of vegetation response in many fields throughout the growing season gives the Tasseled Cap shape. Tasseled Cap is sensitive to light and atmospheric conditions: these parameters must therefore be reduced in pre-processing (Campbell 1996).

4.13 CHAPTER SUMMARY

Meteorological, weather forecasting and vegetation monitoring applications have been discussed in detail in this chapter, as these are fundamental to our understanding of other environmental parameters.

- Nephanalysis uses methods based on the fact that high cold cloud is most likely to be rain-producing cloud and will therefore appear as the coldest cloud on thermal infrared imagery and the whitest on visible imagery.
- However, other parameters observable on satellite imagery are considered important in nephanalysis

such as texture, size, shape, organisation and shadow effects.
- Clustering bi- and tri-spectral plots are considered the most successful techniques for cloud-cover mapping and analysis. They can also provide a useful input to rainfall estimation procedures.
- Rainfall monitoring and estimation techniques can be divided into three major types depending upon their mode of operation: manual, interactive or automatic.
- Rainfall techniques can also be categorised by the mode of analysis that they adopt: cloud-indexing, climatological, life history, bi-spectral, cloud physics or based on passive or active microwave image analysis.
- Automatic techniques tend to be the most widely utilised today and are often best suited for the rainfall in the global environment on which they were first developed. They also tend to be best adapted for either long-period (10 days or more) or short-period (hourly or daily) rainfall estimation.
- Remotely sensed signatures obtained over land often have a vegetation and soil component.
- Ratioing can be used to emphasise vegetation and a number of specialised ratios termed vegetation indices have been developed. These have a wide range of applications: spatial distribution analysis, event detection, Leaf Area Index, fractional vegetation cover, stocks and fluxes of carbon and detection of recent rainfall events.
- The interpretation of crop growth cycles is discussed in conjunction with the crop calendar and Tasseled Cap transformation enhancement.
- Rainfall estimation and vegetation response monitoring are essential to crop and land cover management in cultivated and grazed regions of the Earth's surface. Success or failure of food and economic crops in many parts of the world determines not only local and national ecomomies but also the viability of human settlements in many parts of the world. Any mechanism that enables early warning of such fluctuations in crop status is essential, if not life-saving in many communities. The analysis of satellite imagery often provides the

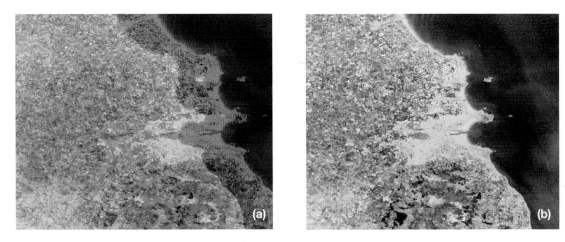

Plate 2.1 (a) False colour MSS image showing offset of band 1
which is projected in red. (b) Corrected false colour image in which the three bands are co-registered.

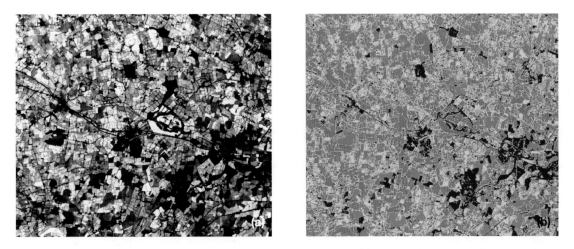

Plate 3.1 (a) TM 4 image shown in its usual mode of varying shades of grey and (b) in density-sliced format. See Table
3.2 for DN bands and their associated colours. Width of image 19 km.

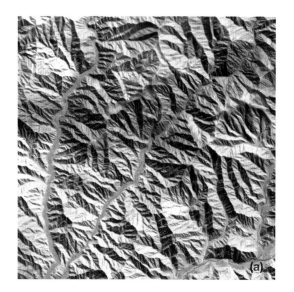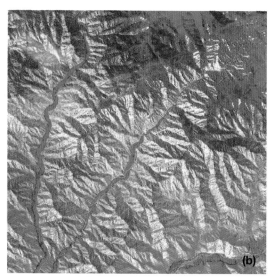

Plate 3.2 (a) False colour image of part of the Andes, Peru produced by projecting TM 5 in red, TM 4 in green and TM 3 in blue. There is no evidence for mineralisation when this band combination is used. (b) A ratio image produced by projecting bands 1/2 in blue, 3/4 in green and 5/7 in red clearly defines an elliptical ore body in the northwest, shown in orange. Area approximately 15 × 15 km. Data courtesy of Neil Quarmby of IS Limited.

Plate 3.3 (a) False colour TM composite of part of Eritrea formed by projecting TM 4 in red, TM 3 in green and TM 2 in blue. (b) Intensity, Hue, Saturation image of the same region. Width approximately 14 km.

Plate 3.4 False colour composite showing different types of vegetation and land use in western Ireland. The image is produced by projecting TM 4 in red, TM 3 in green and TM 5 in blue. Compare with principal component images in Plate 3.5. Area approximately 21 × 21 km.

Plate 3.5 Principal component versions of Plate 3.4. (a) PC1, PC2 and PC3 projected in red, green and blue; (b) PC2, PC3 and PC4 projected in red, green and blue.

Plate 3.6 Principal component version of Plate 3.2a. The image is considerably more colourful than the conventional FCC and the ore body is quite prominent.

Plate 3.7 Synergistic image produced by projecting TM 2 in red, the TM 3/TM 4 ratio in green and ERS-1 radar data in blue. Digital imagery provided by Space Imaging EOSAT Inc.

(a)

(b)

Plate 3.8 Comparison of (a) TM image with (b) TM/SPOT fused data for an airport southeast of Los Angeles. The fused image is considerable sharper than the standard TM image. Digital imagery provided by Space Imaging EOSAT Inc.

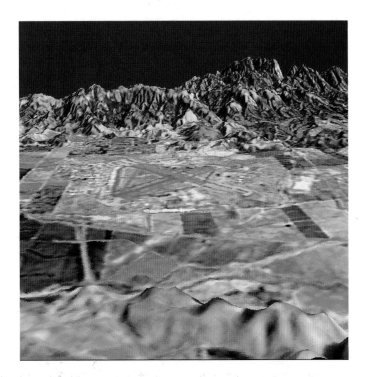

Plate 3.9 Perspective view southeast of Los Angeles produced by draping TM and radar data over a digital elevation model and viewing from the southwest. Digital imagery provided by Space Imaging EOSAT Inc.

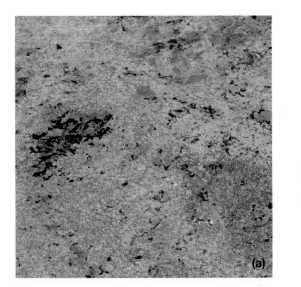

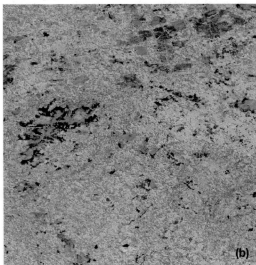

Plate 3.10 Two false colour composite MSS images obtained on (a) 6 May 1989 and (b) 2 May 1990, illustrating the spectral reflectivities on these dates.

Plate 3.11 Density-sliced MSS 4 (infrared) image produced by subtracting the MSS image obtained on 6 May 1989 from the one acquired on 2 May 1990. See Table 3.7 for colour code. Width approximately 70 km.

 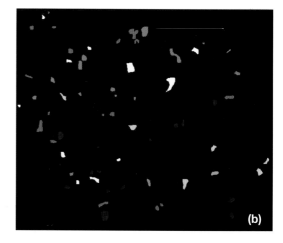

Plate 3.12 (a) False colour composite produced using TM 4 projected in red, TM 5 projected in green and TM 3 projected in blue. Data courtesy of ERAMaptec. (b) Training areas for supervised classification (background image blacked out). See Table 3.8 for colour assigned to each training class.

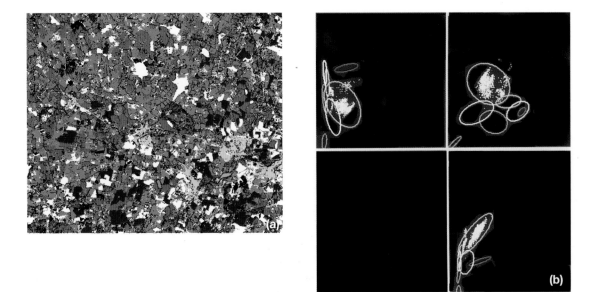

Plate 3.13 (a) Supervised classification representation of Plate 3.12a. Colour assignation shown in Table 3.8.
(b) Feature/space plots for three two-band combinations (clockwise from top left: TM 4/TM5; TM 4/TM 3; TM 5/TM3).
Data for 'soils' shown as white 'cloud'.

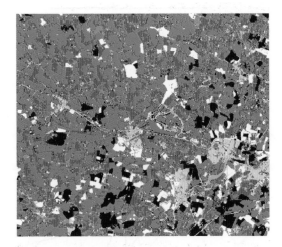

Plate 3.14 Unsupervised classification version of Plate
3.12a using seven classes.

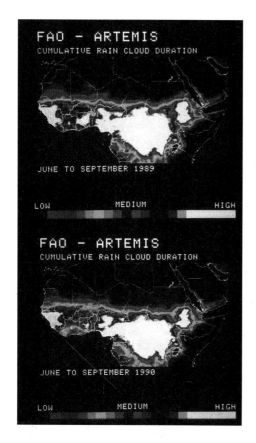

Plate 4.1 ARTEMIS rainfall and cloud product showing cumulative rain-cloud duration for tropical Africa for the period June–September 1989 and 1990. Courtesy of UN FAO Environment and Natural Resources Service.

Plate 4.2 Normalised Difference Vegetation Index of Europe: an example of a global vegetation map. The mosaic of products has been falsely coloured to emphasise variations in vegetation over Europe. Courtesy of NRSC.

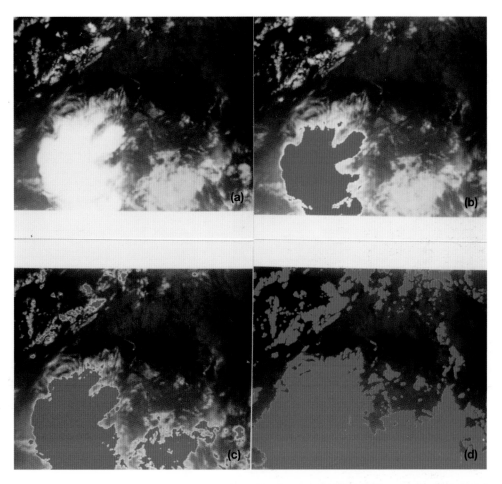

Plate 5.1 Sample of cloud type map for 1 July 1985 (0330). (a) Original thermal image for African Sahel;. (b) High cloud shown in red; (c) High and Middle level cloud shown in red; (d) cloud regions displayed in red, non-cloud in dark blue. (Source C. H. Power). Original data ESA/EUMETSAT 1985.

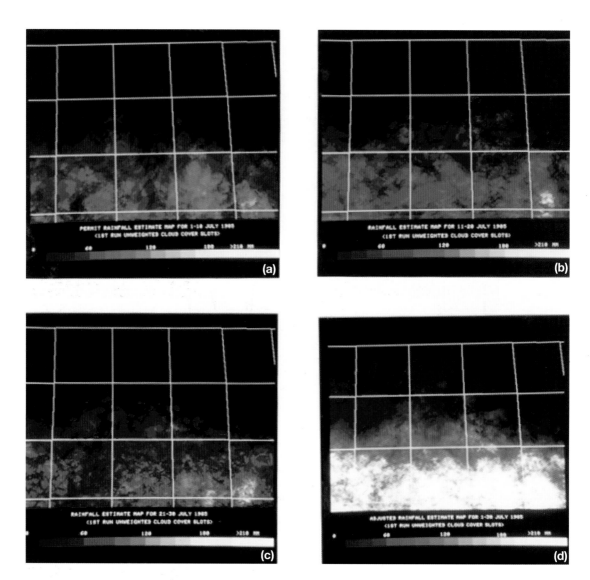

Plate 5.2 Rainfall estimates for July 1985 for 10-day and 30-day periods. Original data ESA/EUMETSAT 1985.

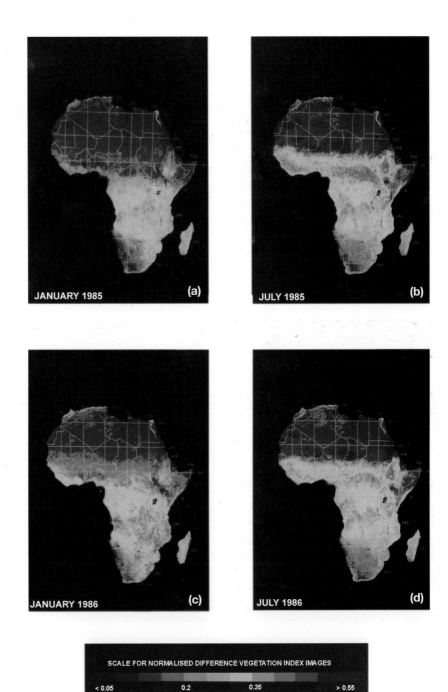

JANUARY 1985 (a)

JULY 1985 (b)

JANUARY 1986 (c)

JULY 1986 (d)

SCALE FOR NORMALISED DIFFERENCE VEGETATION INDEX IMAGES

< 0.05   0.2   0.35   > 0.55

Plate 5.3 Normalised Difference Vegetation Index composite maps for January and July 1985 and 1986: (a) January 1985; (b) July 1985; (c) January 1986; (d) July 1986. Source C. H. Power, courtesy of Remote Sensing Unit, University of Bristol. Original data NOAA 1985/6.

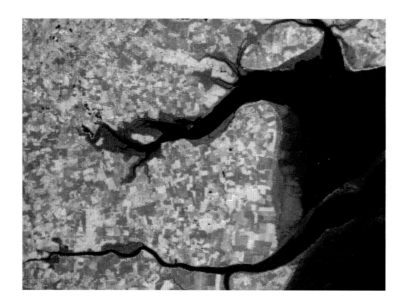

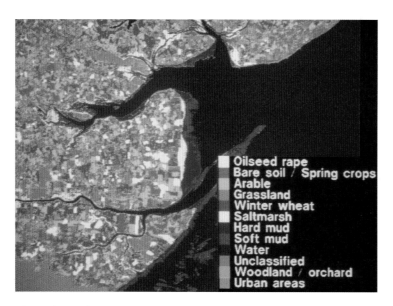

▓	Oilseed rape
	Bare soil / Spring crops
	Arable
	Grassland
	Winter wheat
▓	Saltmarsh
	Hard mud
	Soft mud
	Water
	Unclassified
	Woodland / orchard
	Urban areas

Plate 5.4 (a) A false colour composite of study area with band 3 projected in green, band 4 in red and band 5 in blue. (b) Supervised maximum likelihood classification of land cover on the Essex coast. Copyright NRSC.

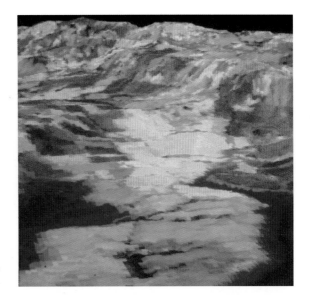

Plate 5.5 The surface model created by draping part of the Landsat TM image over the DEM. Copyright NRSC.

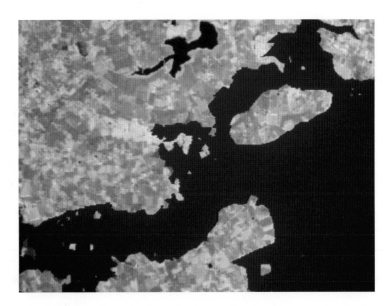

Plate 5.6 Effect of a sea-level rise simulation of 2 m on the northern part of Landsat TM 3, 4, 5 image. Copyright NRSC.

Plate 5.7 An example of peatland ground cover, Wedholme Flow, Cumbria. Copyright Lancaster University Archaeological Unit.

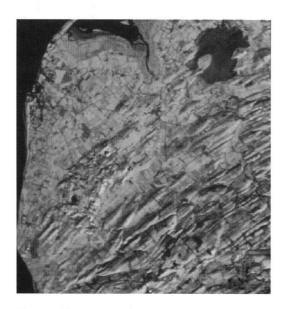

Plate 5.8 The 25 November 1989 Landsat TM scene of the southern part of the area in false colour. Original data from ESA.

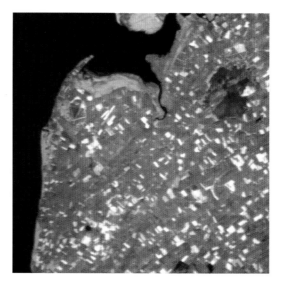

Plate 5.9 A combination of the ratio of band 4 with principal component 4, PC1 and band 5 applied to 26 April 1987 Landsat image of Cumbria proved very effective at highlighting possible peat areas. Copyright NRSC.

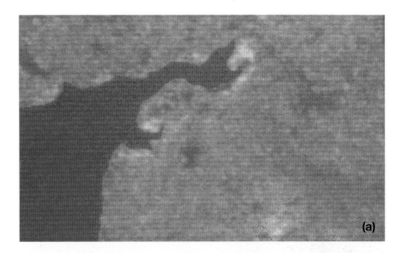

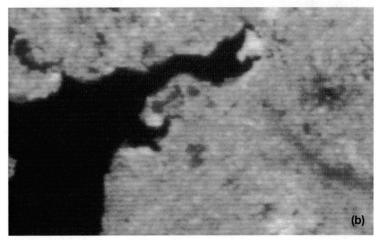

Plate 5.10 (a) PC1 and TM 5 combined with the Hue component showing peat in speckled yellow. (b) Intensity, Hue and Saturation composite showing pear in speckled lilac. Copyright NRSC.

Plate 5.11 As an aid to the visual classification of the study sites the result of the digital classification was used as a 'mask'; where the classified peat category was a single colour (magenta) with surrounding features displayed as the original band data. An automatic classification of all other land uses and features in the Landsat TM image except peat areas is then reversed to produce the final hardcopy product. Copyright NRSC.

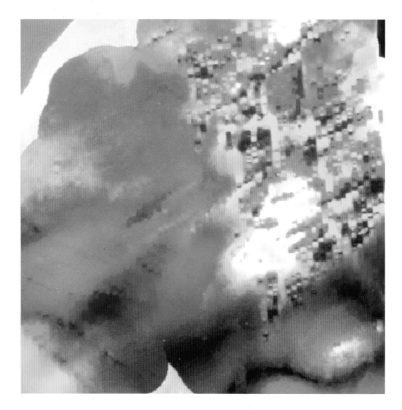

Plate 5.12 Synergistic display of northern Ireland gravity, aeromagnetic and aeromagnetic/gravity data following a principal components transform. See Figure 5.15 for interpretation. Width 130 km. Image courtesy of Taylor and Francis.

only means by which the parameters of rainfall and vegetation can be adequately assessed in spatial extent and timely fashion.

SELF-ASSESSMENT TEST

1 Give the six criteria essential for successful cloud identification on satellite imagery.

2 How can rain-clouds be differentiated from other clouds on visible and thermal infrared satellite images?

3 Give three ways in which the addition of satellite images can improve the assessment of cloud cover and rainfall.

4 Why is low spatial resolution satellite imagery better suited to weather system analysis than high-resolution imagery?

5 What are (a) manual, (b) interactive, (c) automatic rainfall estimation techniques?

6 What are the important spectral characteristics which enable the separation of healthy vegetation from other Earth surface materials?

7 Which band ratio is the best for enhancing differences in vegetation cover?

8 What are the basic assumptions made for vegetation indices?

9 What is a crop calendar and how can it be used to aid satellite image interpretation?

FURTHER READING

Part I: Meteorology and Weather Forecasting

Austin, G. L. and Bellon, A. (1982) 'Very short range forecasting of precipitation by the objective extrapolation of radar and satellite data' in K. A. Browning, (ed.) *Nowcasting*, London: Academic Press, pp. 177–90.

Barrett, E. C. (1987) 'Estimation of precipitation from AVHRR and Meteosat data over Africa', in *Proceedings of the Twentieth International Symposium on Remote Sensing of the Environment, Nairobi, Kenya, 4–10 December 1986*.

Barrett, E. C. and D'Souza, G. (1988). 'A comparative study of candidate techniques for U.S. rainfall monitoring operations using meteorological satellite data', Final Report to the US Department of Commerce under Co-operative Agreement no. NA86AA-H-RA001 (Amendment no. 3).

Bonifacio, R. (1991) 'Rainfall estimation in Africa using remote sensing techniques', in A. S. Belward and C. R. Valenzuela (eds) *Remote Sensing and Geographical Information Systems for Resource Management in Developing Countries*, Kluwer Academic Publishers, pp. 215–33.

Browning, K. A. (1979) 'The FRONTIERS plan: a strategy for using radar and satellite imagery for very short-range precipitation forecasting', *Meteorological Magazine*, 108: 160–84.

Liljas, E. (1982) 'Automated technique for the analysis of satellite cloud imagery', in K. A. Browning (ed.) *Nowcasting*, London: Academic Press, pp. 167–76.

Snijders, F. L. and Minamiguchi, N. (1998) 'Large area monitoring of crop growing conditions by FAO – ARTEMIS', in *Proceedings of International Symposium on Satellite Remote Sensing for the Earth Sciences, Tokyo, Japan, 5–6 March 1998*.

UK Meteorological Office (1997) 'Nimrod: the leading edge of nowcasting' The Met. Office Crown copyright 1997 (97/867).

Part II: Vegetation Monitoring from Satellite

Campbell, J. B. (1996) *Introduction to Remote Sensing*, London: Taylor and Francis.

Kauth, R. J. and Thomas, G. S. (1976) 'The Tasselled Cap, a graphic description of the spectral-temporal development of agricultural crops as seen by Landsat', in Proceedings of Symposium on Machine Processing of Remotely sensed data. West Lafayette, Purdue University.

Pinty, B. and Verstraete, M. M. (1991) 'Extracting information on surface properties from bidirectional reflectance measurements', Journal of Geophysical Research, 96: 2865–74.

Ray, T. W. (1996) 'Frequently-asked questions (FAQ) on vegetation indices and vegetation in remote sensing' available on the Internet: *terrill@mars1.gps.caltech.edu*, created by Terrill W. Ray, Division of Geological and Planetary Sciences, California Institute of Technology, Mail code 170–25, Pasadena, CA 91125, USA.

Richardson, A. J. and Wiegand, C. L. 1977: 'Distinguishing vegetation from soil background', *Photogrammetric Engineering and Remote Sensing*, 43: 1541–52.

5

CASE STUDIES

5.1 INTRODUCTION

This chapter aims to provide the reader with some examples of the use of satellite remote sensing in projects that have been mainly carried out for commercial clients and/or government bodies for 'real-world' environmental applications. It demonstrates how image processing and analysis techniques, such as those described in Chapter 3, can be combined and integrated in order that environmental parameters and phenomena can be better understood. It is designed to show the reader the diversity of effective applications of remote sensing with the purpose of stimulating ideas for independent project work.

5.2 RAINFALL MONITORING AND ASSESSMENT IN WEST AFRICA

Introduction

During the middle and latter half of the twentieth century, Sahelian rainfall totals have been decreasing. There has been neither a consistent downward trend nor any simple pattern in the changing distribution of observed rainfall. However, mean monthly rainfall values for a 50-station series over the Sahel have been shown to be above the 1931 to 1960 mean during the 1950s, but significantly below it during the 1970s. Since 1968 this generally decreasing rainfall over the Sahel has been termed the Sub-Saharan Drought

(Dennett *et al.* 1985). In the mid-1980s, the Sahelian countries suffered from severe repercussions of this drought process. Therefore it is essential to assess the climatic and meteorological processes directly related to these changes in rainfall amounts through rigorous monitoring of both rain-cloud cover and vegetation-cover response. Meteorological satellite images provides a suitable data source for this purpose.

Satellite rainfall assessment and monitoring is a much more difficult task than satellite cloud evaluation, not least because cloud which is bright (cold) in the infrared, and/or cloud which is bright (highly reflective) in the visible, is not always precipitating cloud. Furthermore, particularly in dry environments, precipitation often evaporates before it reaches the ground. Verification of satellite rainfall estimates is difficult in West Africa, for, in the absence of weather radar, ground data are from a sparse network of rain-gauges at point locations only (Figure 5.1). The satellite estimates are for substantial spatial areas, individual image pixels being of the order of 2.5 km^2. To make matters worse, ground-station data are often incomplete and/or unreliable.

The aim of this project was to study the feasibility and reliability of the mapping and monitoring of cloud cover, rainfall and vegetation over the Sahel by satellite. The objective was to study two months during the period when the annual maximum rainfall averages usually occur, namely July and August, and to demonstrate the viability of a satellite-based technique for cloud-cover monitoring,

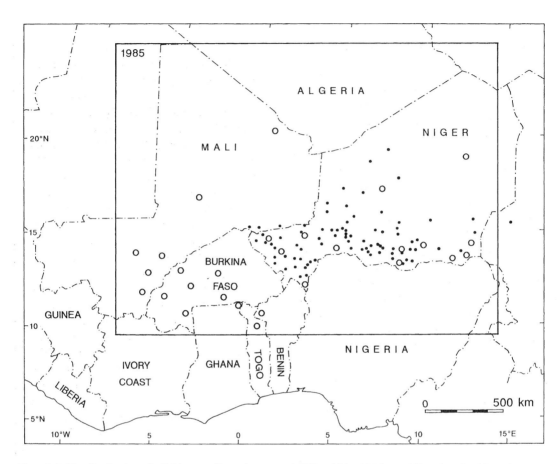

Figure 5.1 The distribution of reliably recording rain-gauges in West Africa 1985, providing data for this project. o = calibration stations and . verifications stations.

and dependent techniques for the evaluation and monitoring of rainfall. The results of this project, based on the infrared and visible cloud imagery of 1985, were then compared to vegetation index images from the 1986 growing season to see whether vegetation response can be predicted from the analysis of rain season imagery.

Methodology

Imagery and Pre-processing Procedures Used for the 1985 Study

Pre-processing of the raw Meteosat imagery from the European Space Agency involved georegistration,

greyscale inversion so that cloud cover would be represented on the image processing software by shades of light grey, and rotations through 180° to facilitate geographic gridding and interpretation. The study area was also subset from the full Meteosat scenes to reduce data-processing volumes.

Cloud types over West Africa tend to belong to cumuliform and cumulonimbus families, but the frequent occurrence of cirriform cloud, commonly in the form of anvils above convectional clouds, presents particular problems for image interpretation. This is because such clouds can be both bright in the visible and cold in the infrared, and thus look like significant rain-clouds despite producing little or no rain at all (Bunting and Hardy 1984). This is where shape of

Climate of the Western Sahel

Over the western Sahel, the annual rain season takes place during the months of July and August when the Intertropical Convergence Zone is in its most northerly position over western Africa and a high proportion of the annual total rainfall is normally expected. Moist onshore winds from the Atlantic Ocean create convectively unstable conditions over much of the region, leading to outbreaks of heavy rain either from relatively well-organised systems or from more scattered cumulonimbus cloud clusters. The best known and most frequent of the organised systems are the so-called 'West African Squall Lines', now known to be associated with easterly wave patterns which move across the region from east to west commonly once every few days. In some parts of the western Sahel two-thirds of the season's precipitation may be associated with these systems. However, along the Guinea coast itself, local rainfall predominately consists of heavy drizzle resulting from very moist onshore air streams. In the north of the region the principal rainfall events are associated with outbreaks of relatively rare and scattered thunderstorm activity.

The highest mean annual and summer seasonal rainfall totals are found along the southwestern and southern coasts of West Africa. Orographic influences, especially over the Futajalon Highlands of northwest Liberia and the Guinea Republic, the Jos Plateau and the Togoland Atacora mountains, locally increase amounts inland. North of 10° latitude, the rainfall distribution pattern becomes more simply zonal (Ojo 1977), becoming drier with progression northwards.

Patterns of mean annual and mean monthly rainfall, raindays and rainfall amounts generally decrease inland from the southwest coast. Mean rainfall and mean numbers of raindays, climatic statistics become increasingly unreliable, the further north one goes from the West African coast towards the fringes of the Sahara. Reliable rain-gauge data are difficult to obtain as one moves away from the more densely populated areas of West Africa, and into sparsely inhabited zones influenced by erratic outbreaks of rainfall. The most intense and more frequent falls of rain are normally expected between 1400 and 1800 local time. In some areas, local topography and wind conditions may alter this pattern so that most rain is likely to occur during the night or into the morning.

cloud formation and weather system identification are important.

Assessments of cloud cover were made over the western Sahel for 10-day and monthly periods. A simple thresholding/density-slicing technique of the kind based on well-established temperature/brightness relationships on thermal infrared images, as described in Chapter 4, was used. The aim was to produce maps of the area of the western Sahel between 9°N–25°N and 7°W–14°E in which the images were classified into four cloud type categories, namely High, Middle, Low and No Cloud. From these, the percentage cover of each class could be determined, and cloud/no-cloud maps created. This approach was applied to four daily Meteosat imaging slots, 0330, 0900, 1500 and 2130. The choice of the slot times was based on the characteristics of the typical diurnal weather cycle found over the Sahel (Ojo 1977). Results from the four daily slots could then be summed and reclassified to create aggregate maps of cloud/no-cloud distributions for each day, for successive 10-day periods, and for the months of July and August as a whole. It was also possible to prepare aggregated cloud/no-cloud maps for July and August 1985 for individual time slots, i.e. for 0330, 0900, 1500 and 2130.

Figure 5.2 illustrates the cloud-cover mapping procedure followed. Typical height ranges for Low, Medium and High-level clouds and their respective temperature ranges over the African Sahel during

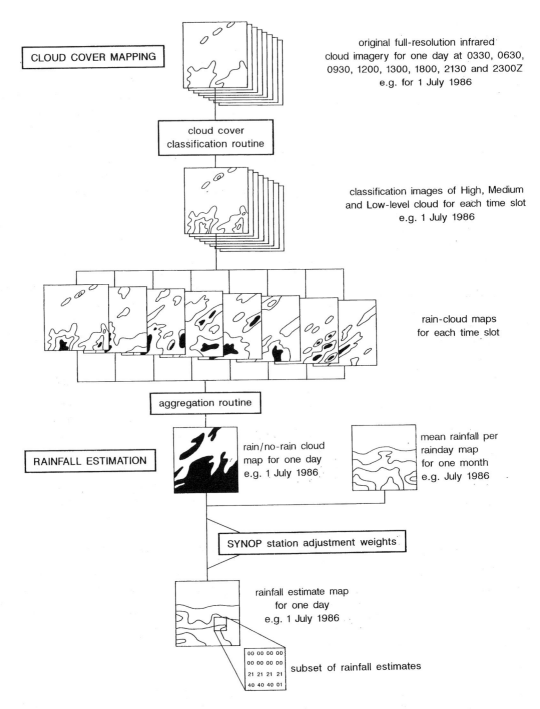

Figure 5.2 Flowchart showing cloud-cover mapping and rainfall estimates procedure (PERMIT) (Polar-orbiter Effective Rainfall Monitoring Integrative Technique).

July and August were selected following consultation of several general climatological texts (Ojo 1977; Barrett 1974; Goldie *et al.* 1958; Griffiths 1972). Vertical sounding charts for West Africa were also examined, and staff of the Foreign and Upper Air Enquiries section of the UK Meteorological Office, Bracknell, were consulted. The consensus is that, within the tropics, the most acceptable distinction between Low and Middle clouds is at about the 500 mb (5,000 m) level, whilst that between Middle and High cloud is at about the 300 mb (9,100 m) level. To separate Low cloud from cloud-free surfaces, a mean minimum ground-level temperature for July and August was derived from published records for the relevant West African countries (Griffiths 1972). Any temperature greater than this value would then be classified as a No Cloud area, and any temperature less than this value would represent cloud cover in one of these three levels selected (Table 5.1). The thresholds between the temperatures were converted to the greyscale brightness responses used on the image processing system using the Meteosat 2 Calibration Report no. 13 for July and August 1985 (ESOC 1985). The resulting greyscale values were inserted into the cloud classification software and tested against a selected sample of images from the total dataset. The processing software was set up to carry out the cloud-cover mapping by density slicing the image at the threshold values selected, and calculating resulting percentages of cloud-cover. Resultant cloud type image maps could be either displayed or printed out.

Results

Examples of the results of the cloud cover analysis for July are shown in Figure 5.3 and Plate 5.1. They illustrate the various stages of density slicing and classification of the original infrared imagery that are entailed in the cloud-cover mapping procedure. Tables 5.2–5.4 summarise some of these results in quantitative form. Table 5.2 presents cloud category and total cloud/no-cloud percentages for a sample day, 1 July 1985. The final (right-hand) column in Table 5.2 gives the percentage of the study region, which remained cloud free during the entire day. Tables 5.3–5.4 present daily cloud/no-cloud statistics and aggregates for July 1985.

Figure 5.3 gives examples of cloud/no-cloud maps for 1 July 1985. Maps of incidences of cloud/no-cloud days for 10- and 31-day periods can also be generated if required from these daily image analyses. Maps of this type can be used to explore the diurnal cycle of cloud cover, and to reveal when rainfall is most likely to occur.

Rainfall Estimation and Mapping

For this study, PERMIT (Polar-orbiter Effective Rainfall Monitoring Integrative Technique) was developed to provide maps of aggregated rainfall over selected areas, for general monitoring and inventory purposes. Such a technique might be particularly well suited for application to areas and situations demanding a relatively cheap and direct approach. The PERMIT method illustrated in the Figure 5.2 flow chart

Table 5.1 Cloud category, height, temperature, pressure and brightness relationships for the cloud-cover mapping on the Meteosat infrared imagery

Cloud category	Average altitude of cloud base	Average pressure level of cloud base	Temperature value	Greyscale/brightness value on Meteosat	Image processor value
Average surface temperature July	n/a	n/a	289.5 K	145 +	109
Low-level cloud	3,000 m	700 mb	268–189.5 K	145	110
Medium-level cloud	5,000 m	500 mb	267–242 K	100	155
High-level cloud	9,100 m	Up to 100 mb	241 K	63	192

Table 5.2 Cloud category percentages for each of the four daily timeslots for 1 July 1985

Timeslot	High	Medium	Low	% total cloud cover	% no-cloud cover	% cloud free during whole day
0330	9.7	8.5	32.6	41.8	58.2	
0900	7.9	7.5	14.6	30.0	70.0	36.6
1500	4.2	6.9	15.0	26.1	73.9	
2130	2.7	12.5	26.6	41.8	58.2	

Table 5.3 Cloud/no-cloud percentages for daily aggregated images for July 1985

Date	% no-cloud cover per day	% cloud cover per day	% no-cloud cover for 10-day intervals	% cloud cover for 10-day intervals	% no-cloud cover for whole month	% cloud cover for whole month
1	36.6	63.4				
2	39.5	60.5				
3	32.6	67.4				
4	35.4	64.6				
5	35.7	64.3				
6	43.6	56.4	8.4	91.6		
7	40.6	59.4				
8	35.1	64.9				
9	37.9	62.1				
10	42.4	57.6				
11	29.8	70.2				
12	22.7	77.3				
13	38.7	61.3				
14	35.1	64.9				
15	32.8	67.2				
16	17.0	83.0	2.3	97.7	0.6	99.4
17	24.3	75.7				
18	29.3	70.7				
19	29.0	71.0				
20	24.2	75.6				
21	19.1	80.9				
22	14.7	85.3				
23	17.1	82.9				
24	24.0	76.0				
25	28.2	71.8				
26	31.6	68.4	0.3	99.7		
27	35.3	64.7				
28	47.9	52.1				
29	35.4	64.6				
30	25.4	74.6				
31	28.1	71.9				

Table 5.4 Cloud/no-cloud cover percentages for the whole month: timeslot aggregates for July 1985

Timeslot	% cloud free at all times	% cloud covered at some time
0330	0.3	99.7
0900	8.0	92.0
1500	4.1	95.9
2130	2.2	97.8

has been designed to function with only four infrared data inputs per day; it can therefore be applied not only to selected geostationary satellite image data but also to data from a pair of polar-orbiting satellites if geostationary satellite data are unavailable.

For more accurate rainfall analyses for significant rainfall events other methods such as BIAS (Bristol-NOAA InterActive Scheme) can be used. This is because BIAS will produce more detailed maps of estimated rainfall after more detailed analysis for selected (unusual or extreme) weather situations.

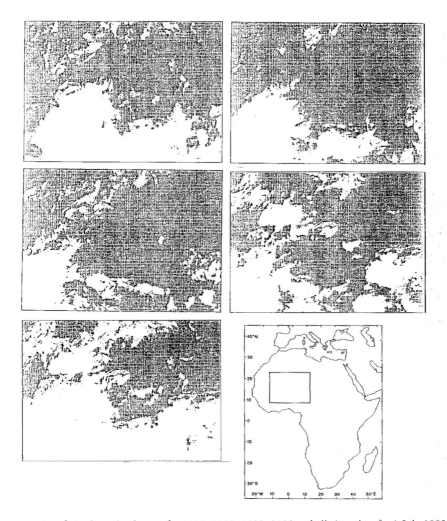

Figure 5.3 Examples of cloud/no-cloud maps for 0330, 0900, 1500, 2130 and all time slots for 1 July 1985 (lower left). White areas represent cloud cover.

With the PERMIT technique, the daily rain-cloud area maps described in the previous section are aggregated to produce 10- and 31-day number of rainday maps. First approximation estimates of rainfall are produced by multiplying the satellite-derived maps of accumulated numbers of raindays by a mean rainfall per rainday statistic derived from climatological atlases (Thompson 1965 and Leroux 1983) (Figure 5.4). Regressions of the first-approximation satellite rainfall estimates on amounts of rainfall observed at selected ground stations are carried out to produce sets of weighting factors to transform the first-approximation satellite rainfall estimates into a final set of satellite rainfall estimates. These are better designed to represent mean 1:1 relationships between the estimates and the ground observations. The purpose of this stage of the technique is to effect a 'current-time' meteorological adjustment to the first-approximation satellite rainfall estimation field, which is calibrated by climatology.

The temperature threshold of the rain-cloud areas adopted was 241 K, i.e. the threshold used in the cloud-cover section as the demarcation between Middle and High cloud. The mean rainfall, mean rainday and mean rainfall amount per rainday, $R \, rd^{-1}$ fields are generated from climatological statistics for the month of July. These were generated from maps in Leroux's (1983) *Climate of Tropical Africa*, which were digitised, co-registered to the Meteosat image projection and transformed into raster or pixel format using standard image processing routines.

Results

Table 5.5 summarises the regression analyses, and Plate 5.2 shows the estimate maps for July 1985 for 10-day and 30-day periods. Although the R^2 values listed in Table 5.5 are modest, the sparseness

Table 5.5 Summaries of regression statistics for observed rainfall reports from major stations in the western Sahel and the equivalent satellite rainfall estimates

Time period	Correlation coefficient	R^2 value (%)
1–10 July	0.61	36.7
11–20 July	0.66	43.5
21–31 July	0.57	32.2

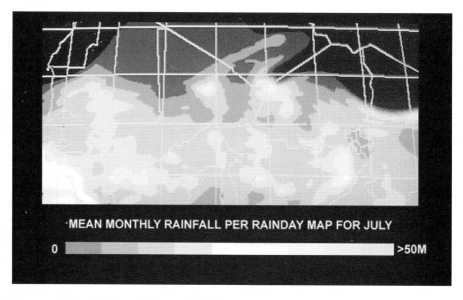

Figure 5.4 Mean monthly rainfall per rainday map for July 1985.

and unreliability of the surface reports suggests that the satellite rainfall maps with which they are associated should generally be much more representative of regional rainfall patterns than comparable maps which might be prepared from surface data alone.

Relationship between 1985 Rain Season Analysis and 1986 Vegetation Growth in the Sahel

Since 1982, NOAA and NASA have produced vegetation indices on a regular basis. The NOAA Environmental Data and Information Service produces a daily Global Vegetation Index from NOAA AVHRR imagery which is retained in a Global Area Coverage database at 4 km resolution. From these, weekly and monthly vegetation composite maps can be prepared. Monthly NDVI (Normalised Difference Vegetation Index) images for mid-season months (January, April, July and October) in Africa were obtained from NOAA so that vegetation distribution changes between the two years could be compared and then related to the July 1985 and 1986 rain seasons in West Africa using the PERMIT rainfall maps. Plate 5.3 shows the NDVI composite maps for January and July 1985 and 1986. Comparison of the two January NDVIs shows that the green vegetation front progressed further north into the Sahara in 1986 than in 1985, which suggests that the 1985 rain season was wetter than the 1984 one.

Conclusions

The PERMIT technique demonstrated how satellite rainfall estimation and nephanalysis techniques can be used to give a better understanding of rainfall events and weather systems. Results from these types of analysis are regularly used by weather forecasters to improve their techniques of weather prediction and to supplement conventional weather report information. The NDVI and PERMIT rainfall map comparisons are only preliminary but they do indicate that there is potential in combining rainfall estimation

from satellite imagery with vegetation assessments to further elucidate seasonal vegetation dynamics in arid and semi-arid regions. These types of analysis are currently contributing to global climate change modelling.

5.3 SEA-LEVEL RISE MODELLING FOR THE WETLAND BIRD HABITATS OF THE ESSEX COAST, ENGLAND

Background and Objectives

This project was carried out for the Wildfowl and Wetlands Advisory Service Ltd by the National Remote Sensing Centre (NRSC) Ltd. The aim of the study is to integrate remotely sensed data with other spatial data using digital elevation data and geographical information systems (GIS) technology for ecological conservation and monitoring applications. The area selected for this study was the Essex coast between Clacton-on-Sea and Southend-on-Sea (Figure 5.5). This coastline encompasses the outfalls of the rivers Crouch, Blackwater and Colne and the mudflats of the Dengie marshes, resulting in a range of coastal and estuarine habitats which attract many wildfowl and wetland bird species. It is therefore an area of great interest to the Wildfowl and Wetlands Trust (WWT).

The Essex coast saltmarshes are threatened by flooding from sea-level rise and by high surge tides estimated to be at a rate of 3–7 mm per year. There has been a substantial loss of saltmarsh in Essex during the last two decades as a result of flooding and high tides. This has happened to such an extent that erosion now threatens the stability and effectiveness of several sea walls as saltmarsh is no longer present to protect the earth walls from erosion. Many of the original earth sea walls constructed after the 1953 flood are in vital need of repair. In many places, declining agricultural profitability and land values are reducing the cost–benefit of repairing the walls, particularly in the marginal areas adjacent to the saltmarshes. The UK government Environmentally Sensitive Areas (ESA) scheme may cause some areas of

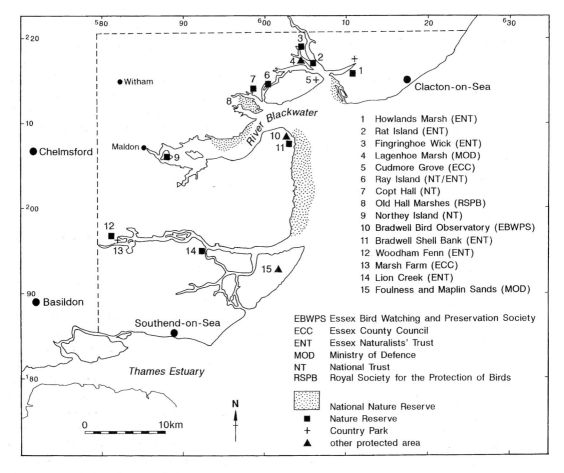

Figure 5.5 Location and extent of the Essex coast study area between Clacton-on-Sea and Southend-on-Sea.

land to revert to extensive unimproved grassland and marshland, a scenario which is favoured for the conservation of the wetland bird habitats. However, in other places, farmers are reluctant to change from intensive crop production, on land reclaimed at considerable expense, to less profitable extensive grazing. Land inside the sea wall is frequently one metre lower than the saltmarsh and the maintenance of crop production will require construction of concrete flood defence walls with sluices enabling rapid drainage in order to prevent crop waterlogging.

The cost of repairing and maintaining the original earth walls is substantially less than replacing them with new concrete ones, and regular inundation of land behind the walls would not only cause land-level rise, with deposition and retention of sediment behind the walls, but also maintain saltmarsh. Other benefits include the provision of recreational areas and better sewage effluent treatment through a natural wetland system, rather than the current treatment works. It is therefore important that the areas of land under threat are identified so that the case can be made for their conservation. Satellite imagery and GIS technology in the study have been used to simu-

Importance of the Essex Coast to Wildlife

The Essex coastline area contains important wildlife habitats and nature reserves. The saltmarshes (about 5,000 ha in area) are particularly important for breeding wader and wildfowl bird populations. These saltmarshes comprise about 11 per cent of the national total, and together with those around the Wash and in Kent are of particular interest as very few are grazed and thus they provide an undisturbed area for breeding birds. A wide range of bird species use this habitat throughout the year and it supports both estuarine and terrestrial species of national and international importance, including the redshank, oystercatcher, wigeon and shelduck, so that significant changes in the size or nature of these areas could have major repercussions on the status of the bird populations.

The bird community of the Essex saltmarshes shows a marked seasonal variation. The diversity of species and the numbers utilising this habitat in winter and during migration passage far exceed the numbers and diversity of resident breeding birds. This is for several reasons: saltmarshes are liable to frequent flooding and are therefore unpredictable as a nesting habitat; they offer little vertical structure as a breeding habitat and this is likely to promote a low diversity of breeding birds, and finally, outside the breeding season, other bird species are more able to take advantage of the saltmarsh plant food resources.

spatial datasets to form natural habitat geographical information systems, which can be used to infer likely population concentrations of certain species of birds and animals and monitor their fluctuations with environmental change. However, at the time of the study satellite data had not been combined with digital elevation data and bird habitat information to study the effect of sea-level rise on coastal wetland bird habitats.

Data Sources

Satellite Imagery

For this project a Landsat TM scene, path 201 row 24 for 29 April 1990 was selected. The image was captured within two hours of low-tide level so that differentiation could be made between mudflats and saltmarsh land-cover types. Although the SPOT system offers a superior spatial resolution, the greater spectral resolution offered by the TM sensor was considered to be more appropriate for habitat mapping as the combination of bands 3, 4 and 5 in the visible, near and mid infrared are better for discriminating subtle differences between vegetation types than the combination of visible and near-infrared SPOT bands. (SPOT 4 data were unavailable at the time of the investigation.) A further impediment to the use of SPOT data in this project was that the study area is located on the boundary of two scenes, which would require time-consuming mosaicing.

Image Band Selection

Bands 3, 4 and 5 were selected as they enhance differences in vegetation types; bands 3 and 4 enhanced differences in water depth and the mudflats; and bands 4 and 5 urban areas and saltmarsh drainage channels. A false colour composite was made with band 3 projected in green, band 4 in red and band 5 in blue (Plate 5.4a). This best enhanced the differences between the classes of land cover (habitats) which the Wildfowl and Wetlands Trust were interested in: namely, mudflat, saltmarsh, grassland, bare

late the effect of sea-level rise on the land use of the Essex Marshes. Landsat TM imagery is most useful in mapping natural habitat land-cover types. Satellite data have been successfully combined with other

soil/young crops, arable crops, winter wheat, oilseed rape, settlements and woodland.

Field Data

The field data included information of land cover in the study area in April 1990, photographs of the wetland areas taken from a microlight and information on bird populations in various habitats in the study area. The land-use data comprised crop-cover data obtained from local farm plans for April 1990, and observations of habitat types (for non-agricultural land) provided by the Essex Project Officer on field maps plotted with Ordnance Survey grid references.

Digital Elevation Data

At the time of the study, the Ordnance Survey offered one standard digital elevation product, a 1:50,000 digital elevation model (DEM), available as either vector format contours, spot heights and break lines or a raster terrain model. This terrain model consists of height values at each intersection of a 50-m horizontal grid. The height values have been rounded to the nearest metre. The accuracy for this dataset is stated by the Ordnance Survey as typically better than 3 m root mean square (RMS) error, with known results ranging from 2 m in hilly rural areas to 3 m in an urban lowland area.

Bird Population Data

The bird population data proved more difficult to obtain in a form suitable for mapping. Enthusiasts obtain many records of bird populations from casual observations during their spare time, so that there is no consistent method or rigour to data collection. The records provided summaries of birds sighted, prepared by organisations like the Essex Bird Watching and Preservation Society and published in field guides. One fundamental problem with collecting reliable bird population data for individual sites is that birds are very mobile and cannot always be asso-ciated with single habitat locations and are often sighted in several different habitats.

Methodology

Image Processing

The imagery was subset to the study area and geo-correction of imagery to the Ordnance Survey National Grid was performed. The 3-, 4-, 5-band composite extract of the Landsat TM scene was classified for land-use/habitat areas using a maximum likelihood classification with ERDAS image-processing software. This classification was then checked against independent field data provided by the Wildlife and Wetlands Trust and was found to be accurate when compared to information in the 1990 farm plans. However, a few modifications were necessary which were then incorporated into the training data of an improved second supervised classification (Plate 5.4b). Subsequently both the original composite and the classified images were combined with the digital elevation model to give a 3D-relief impression of the study area and its wetland bird habitats.

Verification with Field Data

Crop data obtained from local farm records were provided with grid references. These were located on the Landsat TM image and the coloured classes on the image were related to crop types on the farm records. The accuracy of land-cover type/habitat classification was then checked using the remaining WWT ground data not used for the training area selection. This led to the amalgamation of ancient and improved grass-land classes into a grassland class and the barley class became part of the arable crops class as the differences between the subclasses were too subtle for reliable classification. Urban and settlement areas also provided problems with misclassification as settlement areas are made up of small areas of many different land-use types which confuse the classification. It was therefore decided that these areas should be digitised and masked out and the image reclassified without these areas. A second improved supervised classification could then be performed.

Manipulation of Elevation Data

Software routines in the ERDAS image processing package were used to match the elevation data to the image data. It was necessary to extract a section of the data, which matched the study area, and then to resample the data to change the spatial resolution of the elevation data from 50 m to the 30 m of the Landsat imagery. The resampling was done by the cubic convolution method. The composite image of bands 3, 4 and 5 of the Landsat TM image of the Essex coastal area was combined initially with a digital elevation model in order to construct a surface model of the study area. The classified habitat map was also combined with the DEM to produce a surface model of the habitat areas, which could then be used to simulate habitat area losses due to various heights of sea-level rise. Plate 5.5 shows the type of surface model created by draping part of the Landsat TM image over the DEM and Plate 5.6 shows how the northern part of Plate 5.4a would appear if it experienced a sea-level rise of 2 m. Land-cover loss statistics are shown in Table 5.6.

Results

Visual Interpretation of Simulation Images

The Essex coast WWT Project Officer considered the visualisation of the habitat areas and simulation of sea-level rise using hardcopy perspective and plan views very useful in supporting the land-cover loss statistics generated during this study.

Plate 5.6 assumes that the sea defence walls are at a height of 2 m or less so that even the land behind the existing sea walls below 2 m is shown as being inundated by the sea. This would be similar to the situation that would occur if no repairs are carried out to the existing sea walls. Large areas of land would be lost including all the present areas of saltmarsh and mudflat and the reclaimed fields on their edges. However, if the sea walls are repaired, sea-water inundation will depend upon the height of the repaired walls. However, neither the visual interpretation nor the statistics can give us an accurate prediction of the degree to which vegetative succession will generate saltmarsh further inland on areas which are at present agricultural, or whether sea-level rise would cause this land to become too wet to be cultivated or used as pasture. It is therefore impossible to predict accu-

Table 5.6 Land-cover areas lost (in hectares) with different heights of sea-level rise

Land cover class	Area lost (ha)					
	5 m rise	4 m rise	3 m rise	2 m rise	1 m rise	Between low and high tides
Oilseed rape	1,371.8	1,285.7	1,140	872.4	346.6	6.7
Bare soil	4,337.9	4,109	3,858.8	3,407.6	2,295.9	645.3
Arable	3,855.7	3,496.3	3,083.2	2,463.3	1,074.7	4.9,
Grassland	4,982.1	4,584	4,122.8	3,309.3	1,596.5	94.3
Winter wheat	5,636.3	5,200.3	4,708.7	3,861.6	1,827.5	27.8
Saltmarsh	3,636.4	3,614.9	3,587.1	3,526	3,219.1	1,220.8
Hard mud	2,630	2,629.5	2,628.9	2,627.7	2,624	2,507.9
Soft mud	4,913.4	4,913.3	4,913.2	4,912.9	4,910.4	4,808.5
Water	66,338.8	66,334.3	66,327.7	66,319.6	66,291.2	66,234.2
Woodland/orchard	1,868	1,732.4	1,587.7	1,364.3	834.8	55.2
Urban areas	481.9	397.1	326	258.6	162.1	10.1
Unclassified	2,705.4	2,588.2	2,443.7	2,192.4	1,556.4	498.1
Total	102,757.7	100,885	98,727.8	95,115.7	86,739.2	76,113.8

rately the full effect that sea-level rise will have on bird habitats. However, this study gives a starting point and visual representation of the worst possible scenario, that all the bird habitats are lost.

Generation of Supporting Statistics

Bird species are very dynamic within saltmarsh and mudflat zones and their numbers vary from season to season and from year to year among the Essex estuaries, so that it is very difficult to assign bird population numbers to specific habitat areas. Nesting numbers will depend on seasonal changes in flooding and tidal patterns. Therefore, no attempt was made to create a digital map of bird populations within habitats. However, by simulating land/habitat loss from sea-level rise, particularly of mudflat and saltmarsh, and by relating this to the areas where key bird species have traditionally nested, implications for the change in future potential breeding areas for bird species can be made, and tentative forecasts as to whether this will reduce breeding and wintering bird populations in the future. To this end statistics of habitat area loss have been generated.

Discussion of Results and Conclusions

Most, if not all, of the saltmarsh in the study area is contained below the 5 m level. A comparison of the high/low tide and 1 m columns in Table 5.6 shows that the greatest amount of saltmarsh is in this zone and would be lost with a 1 m sea-level rise. The land cover in this zone is the most important for the aims of this study, as this is the key habitat area for the wetland bird species. The largest area of tidally exposed hard and soft mud is also at this level, which is an important feeding ground for the wetland and particularly wading bird species.

If the mean sea level of the Essex coast rose permanently by 1 m and the existing sea walls were not repaired, the results indicate that over 3,219 hectares of saltmarsh and 7,534 hectares of mudflat would be lost. However, if this occurred, other land-cover types would probably be modified by more frequent high-

tide level sea-water inundation above the 1 m mark. Thus, what is now viable pasture and cropland may revert to saltmarsh and/or poor grassland by the natural process of plant succession replacing some of the lost saltmarsh habitat. If, however, sea defences are repaired and improved, the land on the seaward side of the defence walls will be exposed only between high tides, so that it is unlikely that saltmarsh will survive. The land behind the sea defence walls is unlikely to receive sufficiently frequent salt-water inundation to remain saltmarsh and land reclamation pressure for agriculture will mean that the saltmarsh will be permanently lost.

It is very difficult to predict whether, even if new saltmarsh is created, the bird species will use it as a habitat or whether future agricultural land use will allow new saltmarsh to generate landward of the sea walls. However, the present study has revealed some of the planning problems which need consideration and demonstrated the worst possible scenarios of the total loss of saltmarsh.

Future Developments

This study shows that useful information for environmental planning and modelling can be gained by combining satellite image data with a DEM. Although it is difficult to create coverage maps of bird population data, particularly in an area where the seasonal variation in bird populations is so great, useful assessments of the likely impact of sea-level rise and the sensitivity of the bird habitats to it can be obtained from the database set up in this study. This information could be used to develop Environmental Sensitivity Index (ESI) maps for the Essex coast bird population habitats according to the value of the land cover as habitat/breeding areas for different species of waders and wildfowl. With the use of GIS modelling, this could be related to human influences such as sea-wall repair and local future land-use plans. If more reliable bird population data could be obtained, mean bird population figures could be associated with specific grid reference locations or habitat areas. The existing database simulations of sea-level rise could be produced according to sea defence wall type and height, i.e. whether the walls would be overtopped by high tides if they were built to certain different heights; from this the frequency of inundation can be used to predict whether certain habitats would remain viable or not under different sea-level rise scenarios. The simulation of sea-level rise would be more accurate with more detailed digital elevation data than are currently available for civilian use. Future improvements in the spatial resolution of satellite altimeter data may improve the situation.

Alternatively, the release of more detailed military DEM data, or a detailed field survey of the coastal area, would also enable the more widespread usage of DEM data with GIS technology in the sphere of environmental planning and modelling.

This study could be extended to create a land-use/habitat sea-level change monitoring GIS. This could be done by extracting land-use and sea-level information from aerial photographs from the 1950s and re-mapping the information using GIS technology so that the data can be cross-referenced with the maps derived from this project and the change in land use/sea level and land reclamation can be seen. This capability would be useful in the future planning of sea defence construction in the Essex coastal area, creating a system similar to that described by Sader *et al.* (1991).

5.4 TROPICAL CYCLONE TRACKING OVER THE ARABIAN SEA AND SULTANATE OF OMAN FROM 1966 TO 1987

Introduction

This case study comprises project work carried out for the CCEWR (Council for Conservation of the Environment and Water Resources) of the Sultanate of Oman in 1989. The aim of the project was to use the complete NOAA 35 mm microfilm satellite image archive covering the Arabian Sea and adjacent areas, from December 1966 to December 1987, to locate and track all the tropical cyclones that occurred in the vicinity of the Sultanate of Oman. This 20-year record (1978 being omitted) was then studied to

determine the individual sizes, characteristics, frequency, seasonality and favoured locations of the cyclogenesis, activity and dissipation of storms.

Methodology

The entire 20-year NOAA global 35 mm microfilm archive of visible, daytime and night-time infrared imagery was scanned by a microfilm reader. The dates of tropical storms and cyclones in the study region bounded by eastings 45° and 80° and northings 0° and 30° were identified. Examples of this imagery are shown in Figure 5.6. The satellite images in the

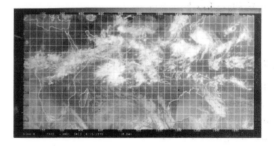

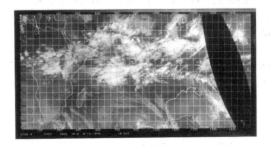

Figure 5.6 Examples of the NOAA global 35 mm microfilm archive of visible and infrared meteorological satellite imagery.

archive cover most of the world and are mapped to both the polar-stereographic and the Mercator projections. Only visible images were available throughout the 20-year period, and the cyclone maps were therefore derived from the visible imagery; the infrared data were used to confirm the presence or absence of a cyclone, or for deriving maps on the rare occasions when the visible image was missing. If a suspected cyclone was recorded, the images for the days immediately before and after the date recorded were reviewed to see whether the storm satisfied the defining cyclone criteria:

1 The storm persisted for two or more days.
2 It exhibited some degree of rotation during its life.
3 The storm cloud was present on all available data for all the days of the cyclone's duration.
4 During its life, the storm was located over coastal or sea/ocean areas for most of its existence.

When a cyclone was located, the cloud-canopy boundary of the storm and the rotation centres of the cyclone cells were also plotted on to Mercator projection base maps. One map per storm was used, annotated with the dates, latitude/longitude locations of the storm centres and a classification for the cyclone cloud canopy for each day of its duration. The classification was based on the World Meteorological Organisation (WMO) scheme for common tropical cyclone patterns and their corresponding T numbers. The T-number categories give an indication of the severity of each storm canopy and were recorded so that subsequent 'area of storm impact' assessments could be made.

The dates and durations of the tropical cyclones were checked against as many conventional records as possible: the European Meteorological Bulletins from 1973 onwards and the Russian Meteorological Bulletins for the period 1966 to 1973 were consulted. A computer search for literature and records of tropical cyclones in the Oman, Arabian Sea and Indian regions was carried out at the National Meteorological Office Library in Bracknell, UK.

This compilation of conventional records is given in Table 5.7 alongside the list derived from the satellite microfilm study. The microfilm-derived listing is considered the more comprehensive and accurate as the conventional listings are by no means consistent in their coverage of every year from 1966 until 1987. They often rely on weather reports from ships or aeroplanes, but these depend upon such craft being in the vicinity of a storm in order to make a record. If a storm does not make landfall, then it is possible that it is never recorded by conventional means. However, satellite image data have the potential to give accurate records of the durations, tracks and extents of these storms day by day. Even if the satellite pass misses the storm on one day, the orbits of such satellites are designed so that they cover the whole globe at least once in a two-day period. Table 5.8 lists the number of days within each year when images were unavailable.

Analysis of the Tropical Cyclone Tracks and Maps

Manually plotted maps of each cyclone track (Figure 5.7a) were digitised onto the computer-stored file of the base map so that the cyclones could be analysed and their distributions quantified. The maps in Figures 5.7b and 5.7c illustrate the variable shapes and natures of the storms. The cyclone tracks for the whole 20-year period were plotted onto two base maps as summaries of cyclones for the total period (Figure 5.8). Computer routines were derived to identify the most common zones of cyclogenesis and cyclone dissipation in the study area. Each storm could also be analysed for 'area of storm impact' by measuring the areas of the individual daily storm cell cloud canopies. From these analyses, likely locations and areas of impact of tropical cyclones were determined, to enable better forecasting in the future. From Figures 5.9a and 5.9b it is possible to locate the areas most frequented by cyclogenesis and cyclone dissipation in the Oman/Arabian Sea region. Finally, a series of statistical bar charts was generated to summarise findings and these are shown in Figure 5.10.

Table 5.7 Comparison of number of tropical cyclones and severe storms over the Arabian Sea

Year	Satellite record			Conventional record		
	Month	Dates	Duration	Month	Dates	Duration
1967	May	10–15	6			
	June	12–14	3			
		18–22	5			
	Sep	22–29	8			
1968	May	25–27	3			
	Aug	05–07	3			
1969	May	19–22	4			
	May/June	30–07	9			
	June	25–30	6			
	July	20–22	3			
	Aug	16–19	4			
		22–26	5			
	Oct	20–23	4			
	Nov	17–19	3			
		27–29	3			
	Dec	03–09	7			
1970	May/June	28–01	5	May/June	31–04	5
	Aug	07–10	4			
		27–29	3			
	Sep	05–10	6			
	Nov	21–25	5	Nov	19–29	11
1971				Sep	24–28	5
	Oct	27–31	5	Oct/Nov	27–01	6
	Dec	14–19	6			
1972	June/July	29–02	4			
	Oct	22–24	3	Oct	24	1
1973	June	05–10	6			
	Dec	26–28	3			
1974	April	13–17	5	April	12–17	6
	May	17–22	6			
	Sep	19–24	6			
1975	May	01–11	12			
	June	24–26	3	Oct	22	1
1976	NONE SIGHTED			NO RECORDS		
1977	June	08–14	7	June	09–13	5
	June/July	29–02	4			
	Aug	08–10	3			
	Sep	03–06	4			
	Oct	19–22	4	Oct	20	1
	Nov	11–21	11	Nov	12–23	12
1978	IMAGE DATA MISSING			Nov	05–13	9
				Nov	24–29	6
1979				May	12	1
	June	16–22	7	June	16–20	5
	Aug	09–14	6			
				Sep	18–24	7
				Nov	13–17	5

Table 5.7 Continued

Year	Satellite record			Conventional record		
	Month	Dates	Duration	Month	Dates	Duration
1980	June	20, 24–26	7			
	Oct	12–18	7			
	Nov	16–18	3			
1981	Oct/Nov	28–03	7	Oct/Nov	28–03	7
1982	Feb	20–22	3			
	June	17–20	4			
	Sep/Oct	30–02	3			
	Nov	06–09	4	Nov	05–09	5
1983	Aug	08–12	5	Aug	10	1
1984				May	26–28	3
	Dec	03–07	5	Nov/Dec	28–08	11
1985	NONE SIGHTED			NO RECORDS		
1986	June	05–12	8			
	Oct/Nov	31–03	4			
	Nov	07–11	5			
1987	June	06–12	7	June	05–09	5
	Dec	08–11	4			

Note: The full NOAA 35 mm microfilm satellite data archive from 1967 to 1988, with conventional and ship recordings over the same time period have been used

Sources: European Meteorological Bulletin, various journal articles from Mausam (Indian Meteorological Department). This listing has been derived from several different sources and therefore may not be complete as no one continuous source of conventionally recorded cyclones in this region was available at the time of compilation.

Results

The plots of the cyclone tracks in Figure 5.8 show that there are areas favourable to cyclogenesis and cyclone dissipation. The tropical cyclones detected over this 20-year period travel predominantly from east to west across the Arabian Sea. However, a few take curved or even rotational tracks back on themselves. The most common positions for cyclone development tend to be between 70–80°E, 00–10°N; 67–72°E, 15–20°N; and 67–72°E, 20–25°N and decay between 55–65°E and 15–25°N (Figures 5.9a and 5.9b). Many cyclones develop over the sea to the west of India and travel across the Arabian Sea in a northwesterly direction, but they often dissipate between 60 and 65°E, before reaching Oman. Once cyclones make landfall, their storm energy usually weakens and they dissipate within 24–36 hours.

An interesting discovery was made by plotting the cyclone tracks on two separate maps, one from December 1966 to 1970, and the second from January 1971 to December 1987. In general, the cyclones in the earlier period are distributed in tracks parallel to the Indian coast and they rarely travel further west than 60°E, with only two reaching Oman, whereas those in the later period seem to form in the middle of the Arabian Sea and travel west, quite often reaching the coasts of Oman and Saudi Arabia. Analysis of the storm areas of influence from the digital maps shows that in the decade 1977–1987 storms had a larger area of coverage than in the first decade, 1967–1976.

Table 5.8 Image data dropouts over the Arabian Sea from NOAA 35 mm microfilm archive 1967–1987 inclusive

Year	Total number of days missing imagery
1966 (Nov and Dec only)	14
1967	57
1968	80
1969	4
1970	17
1971	62
1972	73
1973	5
1974	16
1975	12
1976	6
1977	1
1978	Whole year missing
1979	31
1980	20
1981	8
1982	9
1983	6
1984	15
1985	81
1986	23
1987	18
Total	558

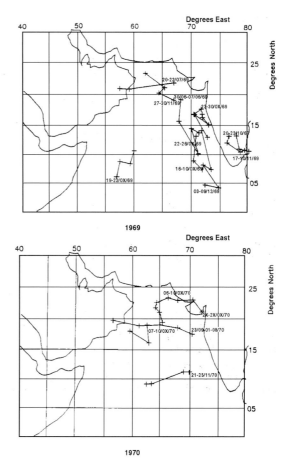

Figure 5.7 (a)

Conclusions

From the summary bar charts in Figure 5.10 some general trends can be found. Most tropical cyclones over this 20-year period have a duration of four to five days (Figure 5.10a). More cyclones occurred in June over the time period studied (Figure 5.10b) and, although some years have more cyclones than others, the average number is three per year over the 20-year period (Figure 5.10c). The years having the highest areal coverage of tropical cyclones in the Arabian Sea region were 1982, 1981, 1977 and 1969 (Figure 5.10d). Over the 20-year period, the number of cyclones has decreased but cyclones tend to have a larger area of influence and travel farther westwards than in the earlier part of the study period. These satellite observations have important implications for the future prediction of flooding and severe weather in the coastal and mountain areas of northern Oman, which would have been difficult, if not impossible, to detect using conventional records.

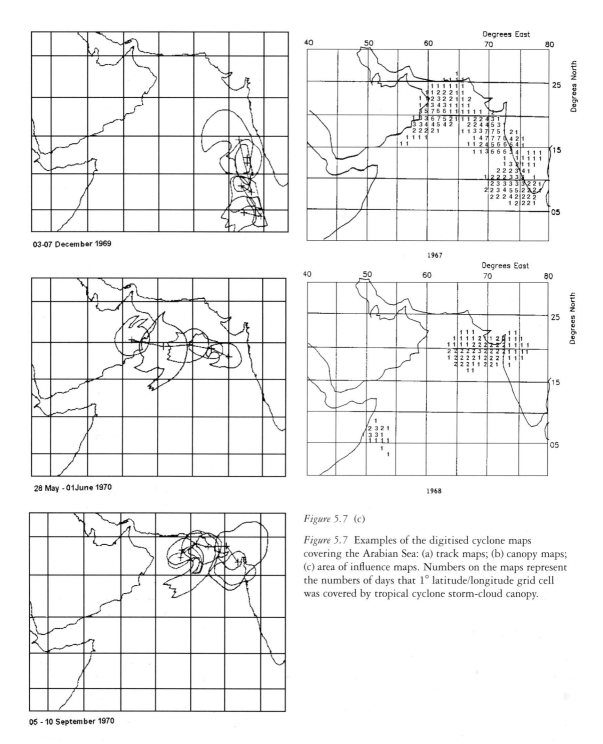

03-07 December 1969

28 May - 01 June 1970

05 - 10 September 1970

Figure 5.7 (b)

Figure 5.7 (c)

Figure 5.7 Examples of the digitised cyclone maps covering the Arabian Sea: (a) track maps; (b) canopy maps; (c) area of influence maps. Numbers on the maps represent the numbers of days that 1° latitude/longitude grid cell was covered by tropical cyclone storm-cloud canopy.

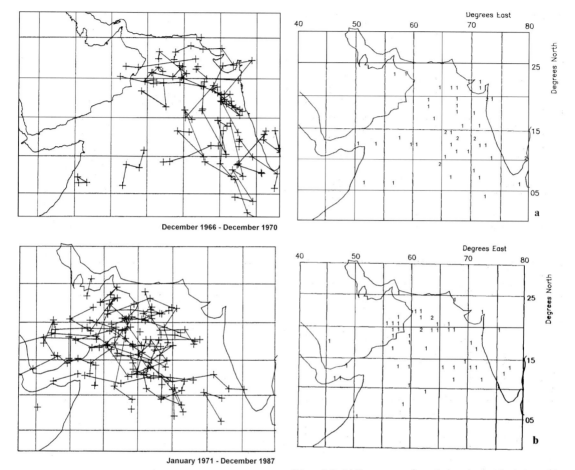

December 1966 - December 1970

January 1971 - December 1987

Figure 5.8 Tracks of tropical cyclones for December 1966 to December 1970 (inclusive) and January 1971 to December 1987 (excluding 1978).

Figure 5.9 (a) Frequency of tropical cyclone genesis (cyclogenesis) 1967–1987. (b) Frequency of cyclone dissipation per 1° latitude/longitude grid square. Numbers refer to number of cyclones.

5.5 DETECTION OF PEAT SOILS FROM SATELLITE IMAGERY

Introduction

This study is part of the North West Wetlands Survey (NWWS), established by English Heritage in 1987 to survey the archaeology and environmental history of the lowland peat of northwest England. The NWWS is funded by English Heritage, and combines documentary and northwest England palaeoenvironmental research with archaeological and environmental fieldwork and aerial archaeology (Middleton 1990). Many valuable archaeological finds have been discovered in peatland areas, well preserved in the permanently waterlogged terrain. Being able to locate areas of peat is therefore of major importance to field archaeologists. The NWWS aimed to produce a management and conservation strategy for the wetlands of northwest England based upon the results of this integrated survey. The areas chosen for this pilot project are shown in Figure 5.11. They contain dif-

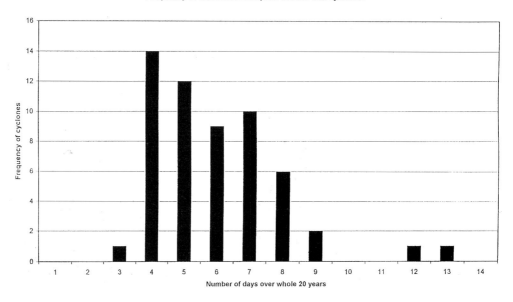

Figure 5.10 Tropical cyclones: (a) Frequency of duration of tropical cyclones.

fering peat distribution, land use and topography and provide good examples of peat types.

The objective of this study was to investigate how satellite imagery and digital image processing could be utilised in the detection of lowland peat areas. The study concentrated on identifying those peatland areas which underlie current agricultural land uses, where the differentiation of soil types is often difficult. In particular, smaller areas of undisturbed lowland peat were targeted as these areas are of particular scientific interest to field archaeologists because they may represent undiscovered sites containing preserved artefacts and information about prehistoric land use.

Little work has been published on the detection of peat from satellites, although some notable work was carried out in the late 1970s and 1980s by the Macaulay Land Use Research Institute. Stove, Hulme and Robertson carried out extensive studies on upland moorland peat on the Isle of Lewis (Stove, Hulme and Robertson 1980; Stove and Hulme 1980), using a combination of ground survey and aerial photogrammetry to test the use of Landsat MSS

imagery for peat detection and classification. The interpretation of the Landsat image revealed certain peatland categories at a much lower spatial resolution but with a greater contrast in ground reflectance than aerial photographs. Landsat MSS band 7 (0.8–1.1 μm) was the most useful for detecting the greatest contrast in response within peat areas, whilst also enhancing hydrological features and delineating wet and dry terrain. Overall, Stove, Hulme and Robertson found that there was a high degree of areal correlation between Landsat classification categories and corresponding photogrammetrically mapped categories. Digital image processing enhancement enabled the detection of different peat-cover types for accurate mapping of image-derived classes to 1:25,000 or larger scales.

Horgan and Critchley (1989) carried out a study to ascertain the value of Landsat TM imagery for peatland mapping and mineral exploration in Ireland. They found that bands 3, 4 and 5 were most suitable for mapping discrete peatlands in areas of mixed land use, and a mathematical morphology technique was

Number of cyclone incidences per month (over 20 year period 1967-1987)

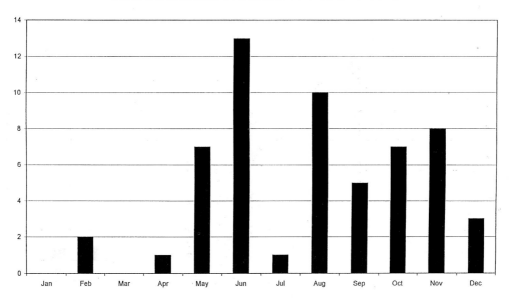

Figure 5.10 (b) Number of tropical cyclone incidences per month over 20-year period.

used to separate peatlands from other types of land cover automatically. Maximum likelihood classification was used to differentiate between various types of peatland (undisturbed, newly cut, machine cut, hand cut and burnt peat). A multitemporal approach was also adopted and it was concluded that the summer image alone or a summer image in combination with winter and spring images are better for distinguishing peat types. This evidence suggests that it is possible to map peat from satellite images but also that more work needs to be carried out, particularly for detailing lowland and mixed peat soils.

Methodology

Peat Discrimination

There were three types of peat evident in the north-west wetlands study area: first, blanket/flow peat, which exists in lowland and upland areas as large intact surface patches, commonly covered by heath vegetation; second, reclaimed peat which was formerly waterlogged or below sea level; and lastly peat

soils which have been modified by agriculture (Knight 1991). Peat cutting generally takes place in the first two categories. This project is concerned with the detection of the latter two categories: reclaimed peat and peat soils. Plate 5.7 shows a typical example of peatland in this region.

Ground Data

Ground data for the study were provided in the form of vegetation and soil observations plotted on a 1:25,000 Ordnance Survey map by the Lancaster University Archaeological Unit (LUAU) and soil survey data plotted onto a 1:50,000 map.

Aerial Photographic Survey

The Cambridge University Committee for Aerial Photography (CUCAP) acquired the aerial photographs used for this assessment. The monochrome prints were from vertical stereoscopic runs covering the majority of the study areas. Colour verticals were also consulted, as film positives, for 4 July 1983 at a

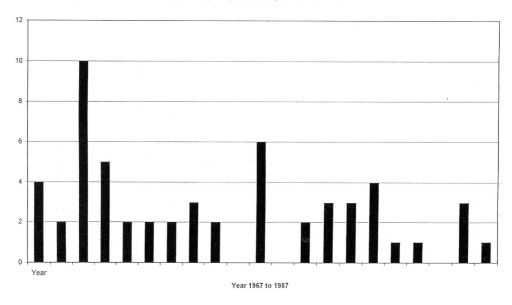

Figure 5.10 (c) Yearly totals of tropical cyclone occurrences.

scale of 1:5,000, covering Scaleby Moss (NY 435638) and Salta Moss (NY 085450).

Photographic Interpretation

Aerial photographic interpretation was carried out according to the principles set out by Carroll *et al.* (1977). All photographs were examined stereoscopically and annotated overlays prepared to indicate the boundaries of possible areas of peat. The initial photographic interpretation was completed without reference to maps, to test the interpreter's ability to recognise the peat deposits and their related topography, land use and general appearance. The three types of peat were identified and differentiated on the aerial photographic coverage. Large areas of intact peat moss, such as those occurring at Wedholme Flow (NY 220530) (Plate 5.7), Solway Moss (NY 350690) and Bowness Common and Glasson Moss (NY 230605), are easily identified by their uncultivated

appearance, which contrasts sharply with surrounding arable and pastoral land uses. Commercial and manual peat cutting is easily identified in these areas. Semi-reclaimed moss and smaller patches of isolated 'bog' (such as that at Bigland Bog, NY 258537) show as differences in land-surface texture that are due to the vegetation types supported by the relatively wet soil (some scrubby areas, tussocky grassland, or mixed, rather stunted deciduous woodland) and the land use (predominantly permanent rough pasture). The 'wet' areas are in many cases respected by existing field boundaries, showing how the 'modern' landscape has been influenced by the natural hydrology and soil of this area (Burgh-by-Sands, NY 330600). Wetland areas are seen to follow the topography very closely, and occur as expected in low-lying locations and in depressions between ridges of slightly higher ground which are easily observed stereoscopically. The colour vertical photographs were consulted to confirm the interpretation of vegetation differences

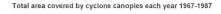

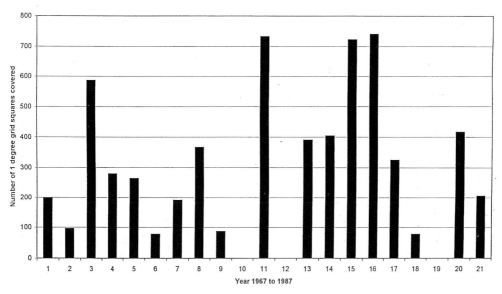

Figure 5.10 (d) Total area covered by cyclone canopies (area in 1° latitude/longitude grid squares).

between preserved peat mosses, semi-reclaimed mosses and fully reclaimed wetland. Areas of peat, which are now fully drained and taken into arable cultivation, are much more difficult to identify on aerial photographs. The drainage method is often evidenced by the marks seen in crops caused by machine-dug drainage trenches. Areas of bare soil are detected by the tonal differences between dark peat soils and surrounding sands or alluvium, and in some cases provide good evidence of peat distribution.

The completed aerial photography interpretations were compared to maps of known wetlands and peat deposits, generally with a good correlation between the two datasets. In some cases, the air photograph data differed from or added to the data mapped by the Ordnance Survey, the soil survey and the NWWS. The results of the photographic interpretations were transcribed to working maps at 1:25,000 scale, outlining areas of probable wetland. These maps were then used to verify the visual interpreta-

tions and classifications of the satellite imagery. Figure 5.12 shows an example of an aerial photograph interpretation.

Satellite Imagery Available

The study was carried out over two areas in Cumbria (Figure 5.11). The Landsat TM and SPOT paths corresponding with these areas are (TM) 204/22 and (SPOT) 22/238 and/or 22/239. Four suitable Landsat TM and two SPOT cloud-free images were available for the study area from the NRSC Ltd archive. These were examined for their suitability for the project. The SPOT imagery was not considered suitable as it did not provide the range of spectral bands in the mid infrared wavebands required for soil and vegetation discrimination. Its greater spatial resolution tended to fragment areas of peat. The Landsat scenes provided spring, summer and winter coverage of the study region, thus making a multitemporal survey possible.

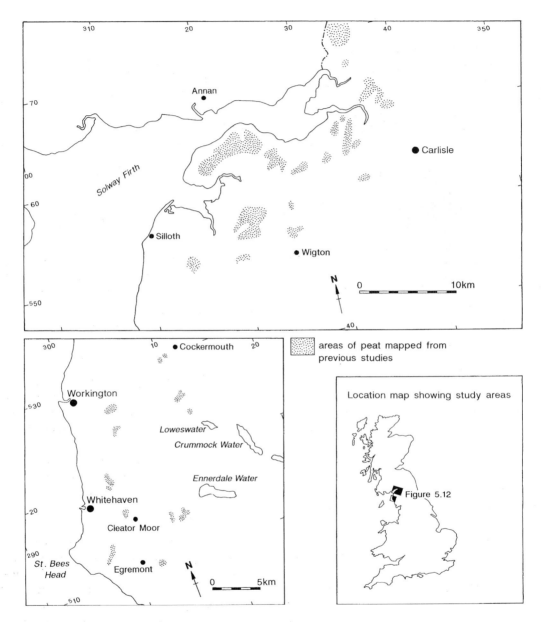

Figure 5.11 Location of study area for peat investigations.

Spectral Separation of Peat

Initial examination of the data rapidly revealed that time of year is an important factor in the identification of peat from TM data. For example, it was difficult to identify the peat clearly on the November scene due to the general wetness and waterlogging of the terrain at this time of year and low sun angle causing terrain shadowing (Plate 5.8). This phenomenon was also observed by Horgan and

 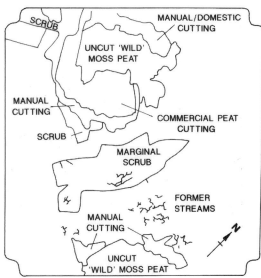

Figure 5.12 Aerial photograph of study area with its interpretation of peat soils. By permission of Cambridge Committee for Aerial Photography. Interpretation by Chris Cox, Air Photography Services, Cambridge.

Critchley (1989). It was also difficult to define peat areas in the July scene, probably for the converse reason that the peat areas had dried out enough to cause confusion with surrounding areas of different soil types. In addition, summer vegetation growth masked and distorted the spectral response of the soil materials in the July image, as observed by Lynn (1984) in an airborne scanner mapping project carried out over the English Fenlands. Although studies by other researchers favour multitemporal approaches to peat classification from satellite imagery, the 26 April 1987 Landsat scene was used for the development of the classification strategy in this study, as it showed the best delineation of peat. TM bands 3, 4, and 5 were considered the most useful for peat classification after inspection and consideration of the band combinations used for peat classification from satellite in previous studies. Commonly, band 3 is used for the discrimination of plant species through chlorophyll absorption, band 4 for the delineation of water bodies and the determination of biomass content and band 5 for the measurement of vegetation and soil moisture. Band 4 was good for peat/sand and coastal

mudflat differentiation and band 5 for drainage and peat cutting, as Stove, Hulme and Robertson found in 1980. Band 5 is particularly useful for measurement of soil moisture and hence is invaluable for detecting peat soils, which retain a high moisture content. However, peat soils underlying agricultural land uses were particularly difficult to differentiate.

Classification of Peat

It was felt that the band combinations found to be suitable for upland peat mapping projects might not be appropriate for lowland peat. New technique development therefore focused on better discrimination of lowland peat areas. Initial analysis of the April TM scene confirmed that conventional automatic classification techniques were unlikely to yield accurate results in terms of identifying total peat area. Whilst the large areas of blanket bog could be easily identified on all the TM bands, other peat types were characterised by a varied spectral response within the individual peat classes; often the variance, also noted by visual inspection, within a class was as great as the variation between classes. The imagery was therefore enhanced to optimise visual interpretation, allowing factors such as texture and context, as well as spectral response, to be easily incorporated into the classification method. Several different band ratios and combinations were tested to produce images where the peat classes could be separated from other land type classes.

Selection of Appropriate Bands for Visual Interpretation

During the visual interpretation it was noticed that the edge of a large peat-flow area (Wedholme Flow) had a response similar to that of sandbank areas on the edge of the Solway Firth. These observations were checked with the field archaeologist, who confirmed that areas of peat erosion and manual peat cutting at the edge of Wedholme Flow had exposed areas of sandy outwash, and that peat had formed on the inland parts of the estuarine mudflats and sandbanks. Although the analysis verified that bands 3, 4 and 5 are the most useful for identifying peat, no single band or band combination could be used successfully to identify all of the areas of peaty soils or reclaimed peat mapped from the aerial photography. Therefore any remaining confusion between these classes was resolved by the ground observations of a field scientist.

Digital Image Enhancements

Given the difficulty of identifying peat from existing TM bands alone, a number of image-enhancement techniques were employed to generate new 'bands' of data upon which peat areas would perhaps be more visible. These techniques fall into two basic categories:

1 Band ratioing
2 Band decorrelation.

In previous studies, ratios have been employed to separate soil types from the overlying vegetation, and to identify different soil-moisture regimes. Therefore several simple band ratios were applied (band 4/band 3, band 5/band 3 and band 5/band 4), along with a number of more complex ratios such as the PVI (Perpendicular Vegetation Index) and the Tasseled Cap technique (Chapter 4). No single ratio was successful in highlighting to a significant degree the smaller areas of lowland peat because of the varied spectral response of the lowland peat types.

Existing bands, and band ratios derived from them, were inherently unsuited to the discrimination of such subtle variations as represented by the change between peat and non-peat areas in the context of a largely agricultural landscape. Principal components analysis (see section 3.7) is a recognised band decorrelation technique which serves to remove the redundant information resulting from the correlation between spectral bands, producing images which give both the dominant and less apparent physical characteristics of an area. Principal component (PC) images were derived from TM bands 1–7. A number of PC and band colour composites were examined for peat discrimination. PCs 1, 3 and 4 were eventually selected for the interpretation as they provided significantly more information on the distribution of peat/damp soils than could be seen in the original data. Dark tones in the PC4 and lighter tones in the PC3 image both closely matched the areas of peat identified from the annotated mapping and aerial photography. However, these composite images tended to overestimate the total amount of 'peat' when compared to the reference sources, probably because they were highlighting areas of wet soil rather than peat, and none of the PC images could be used to differentiate effectively between inland areas of peaty soils and the coastal mudflats and beaches.

In an attempt to resolve the problem of discriminating the peat from the coastal deposits, a ratio was generated using PC4 (the image which gave the greatest amount of information about the distribution of apparent peat areas) and band 4 (the most useful band for separating peat soils from the mudflats/beaches). A combination of this ratio, PC1 and band 5 proved very effective at highlighting the possible peat areas shown in Plate 5.9, where the peat or damp soils appear as pale pink, speckled areas.

Another method of producing new 'bands' of data

is to transform image bands into Intensity, Hue and Saturation (IHS) components as a method of improving the separability of the surface deposits and vegetation elements in the scene (see section 3.6). The three images used to produce the most suitable composite image (band 4/PC4 ratio, PC1 and band 5) were transformed into I, H and S components, and each provided information which could be correlated closely with the available ground-surveyed information. A range of composite images was produced using the I, H and S data in conjunction with either the TM bands or the PC images (see Plates 5.10a and 5.10b). Peat or damp soil areas could be clearly seen on several of these, notably the IHS combination, and upon a PC1, band 5 and Hue composite (Plate 5.10a).

None of the three finally selected composites (band 4/PC4 ratio, PC1 and band 5; PC1, band 5 and Hue; and I, H and S) gave a clear distinction between areas of reclaimed peat/peaty soils and areas of merely damp ground. A further source of confusion is the masking properties of certain crops, even in the April images, where the spectral reflectance of the vegetation totally masks out any response from the underlying soil.

Digital Classification

The varied nature of the peatland, i.e. the different vegetation composites located on peat soils, and the differences in the structure of the soil associated with intensive farming and reclamation practices, results in a heterogeneous spectral response, which may inhibit an automatic classification. Instead of the original Landsat TM bands, the digital classification was performed on those bands deemed most useful following the intensive investigation into appropriate bands for a successful visual interpretation. Band 5, PC1, ratio band 4/PC4, and Hue from the April image were the selected bands for the classification.

Detailed ground-survey data were available as training data for a supervised classification of the two study areas. The peat areas delineated by the aerial photography and the soil survey map were used as training data. Crop and vegetation data were also available from the LUAU field archaeologist.

Initially, a supervised maximum likelihood classification was performed on the peat class alone. Although peat areas are clearly defined by a visual inspection of the imagery, the heterogeneous nature of the soil and crop cover described previously produced a poor peat classification by automatic methods. Visual interpretation showed peat areas as speckled. Although this speckled appearance enhanced visual interpretation, it led to problems when incorporated into the automatic classification procedure. Urban, inland water, mudflat and many areas of non-peat grassland resulted in greatest misidentification with peat, probably because these land covers also have high moisture contents.

Given these poor results, a further supervised classification was performed on surrounding non-peat features only. The confusion in the peat classification had come about because of the difficulty in selecting 'spectrally pure' training areas for peat classes, i.e. the inherent speckle of peat classes resulted in a large range of spectral values being incorporated into a single class.

The classification of all other features was successful, leaving just peat and some mudflat areas indistinguishable from one another. As an aid to the visual classification of the study sites, the result of the digital classification was used as a 'mask', where the classified peat category was a single colour with surrounding features displayed as the original data.

The second automatic classification was therefore an aid to the visual interpretation of the scenes. To this end, hardcopy images were produced from the band/PC ratio combinations which best enhanced the previously known peat areas from ground observation, soil survey maps and visual assessment of aerial photographs. Four hardcopies were produced:

1 Combination of band 5, principal component 1 and the ratio of band 4 with principal component 4 (Plate 5.9)
2 PC1 and band 5 combined with the Hue component (Plate 5.10a)

3 Intensity, Hue and Saturation composite (Plate 5.10b)
4 An automatic classification of all other land uses and features in the Landsat TM image except peat areas. This classification was then reversed to produce the final hardcopy (Plate 5.11).

The classified peat areas from the automatic classification image were determined by means of the ArcInfo and ERDAS system, for statistical cross-comparison with the visual interpretation and the aerial photography results.

Conclusions

After appraisal of all available satellite and aerial photographic imagery for the project, it was found that larger scale and higher resolution SPOT imagery did not necessarily provide better discrimination of peatland areas, as in both cases peat soil areas tended to become fragmented and therefore more difficult to differentiate. It is therefore recommended that 1:15,000 scale aerial photographs and Landsat TM imagery at 1:100,000 scale on hardcopy provide the best combination for determination of peat in the northwest lowland areas of England.

Initial subjective cross-comparison of the results from the satellite visual interpretation and the satellite automatic classification with the aerial photograph-derived maps shows that the peat areas produced in the automatic classification of the band 5, PC1, ratio band 4/PC4 and Hue composite seem to coincide better with the aerial photograph-mapped areas than those defined by visual interpretation of the satellite imagery.

Visual interpretation of satellite imagery can be assisted by the application of several appropriate enhancements to the composite imagery beforehand. However, the automatic classification is more time-consuming, particularly in terms of computer time, than the visual interpretation. It appears that for small areas of peat of only a few hectares or less, the automatic classification may be overestimating the number of locations of these areas and the visual interpretation may therefore be more suitable in these cases.

5.6 REGIONAL GEOLOGICAL MAPPING USING DIGITALLY PROCESSED SYNERGISTIC GEOPHYSICAL DATASETS

Introduction

The use of satellite images, especially since the advent of the unmanned Landsat series in 1972, has shown that geological features can be detected and mapped from space. Such imagery is particularly well suited for regional mapping, providing as it does a synoptic view of the Earth. The increased sophistication in the satellite design has been matched by a parallel improvement in image processing software in order that the information extracted could be maximised. However, despite these advances, one continuing disadvantage of such imagery has been that, while it could provide excellent geological information where exposure was good such as in arid terrain, in temperate terrain, with an extensive vegetation cover, especially one that has been greatly modified by agricultural practices, very little information could be extracted. Consequently, in such vegetated terrain geologists often use geophysical data which are conventionally displayed as contour plots in order to obtain geological information. However, using current digital image processing techniques the data can be converted into geophysical images which allow the extraction of geological information even through extensive vegetation cover (Drury and Walker 1987; Lee et al. 1990).

Study Area

The 130 × 130 km study area in northern Ireland is an ideal location in which to evaluate the potential of geophysical datasets. As approximately 98 per cent of the study area is vegetated (or covered by lakes), conventional satellite imagery is of little use for geological investigations. However, extensive geophysical datasets are available: 10,063 gravity and 19,546 aeromagnetic measurements have been made within the study area. In addition, a wide range of igneous, sedimentary and metamorphic lithologies occur within this region, allowing the technique to be comprehensively evaluated.

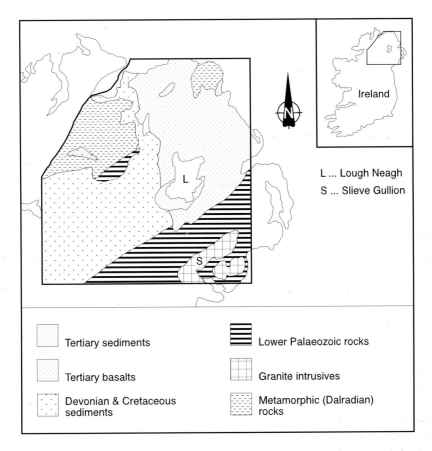

Figure 5.13 Major geological units within the study area.

An inlier of 500 million-year-old Dalradian rocks in northeast Ireland (Figure 5.13), composed mainly of schistose grits and quartzose schists forms the Highland Border Ridge, a continuation of the Dalradian lithologies from Scotland. The largest outcrops of Dalradian are located in the Sperrin mountains to the west and are mainly Upper Dalradian, lithologies composed of quartzite, phyllite and mica and tourmaline schists. Ordovician and Silurian rocks of the Longford Down zone composed mainly of greywackes with associated graptolitic shales crop out to the southeast and are intruded by Caledonian and Tertiary granites. Devonian and Carboniferous sediments crop out extensively in the southwestern sector of the study area. The former consists of conglomerate, sandstone and occasional shale bands whilst the latter consists mainly of clastic sediments overlain by limestones. Fine-grained Tertiary basalts occur over an area $>2,000$ km^2. Post-basaltic sedimentation during the Oligocene was concentrated on the Lough Neagh downwarp, forming thick deposits of sideritic clay, sand and lignite.

Digital Processing

The original gravity (10,063 points) and aeromagnetic (19,546 points) data were converted into a format readable by ERDAS software and the irregularly spaced geophysical data were converted into a regular grid with a pixel size of 260 m in which each pixel was assigned a value dependent on a specified inverse square algorithm, the calculated

Figure 5.14 (a) Gravity image for northern Ireland; (b) aeromagnetic image for northern Ireland. Approximately 130 × 130 km. Images by permission of Taylor and Francis.

distance to the subject pixel and a designated search radius. This manipulation of the data allows each of the two input geophysical data sets to be displayed in a grey-tone format in which the lower the value the darker the image (Figure 5.14). In order to produce a third input band, which is necessary to generate a full-colour image (where each input band is projected in red, green and blue), one band was synthetically formed by using the aeromagnetic/gravity ratio as the third input. The inclusion of the aeromagnetic/gravity ratio in a false colour display results in a relatively bland image because all three input bands are not statistically independent. A principal components transform was performed on the input data and the three calculated principal components were projected in red, green and blue to produce Plate 5.12. This transformation yielded a spheroidal rather than an ellipsoidal data distribution which makes better use of the available colour space. The use of such intense colours optimises the information extraction by the human visual system.

Interpretation of Synergistic Display

Gravity and magnetic images are shown in Figure 5.14. The magnetic image is dominated by Tertiary basalt in the northeast which yields a large variance signature because of the high remanent magnetism of these igneous rocks and a series of subparallel 40 km-long dark lineaments in the southwest. These are reversely magnetised igneous dykes. The bright oval signature in the southeast from a gabbro contrasts well with the low magnetic values trending ENE–WSW over the Lower Palaeozoic greywackes. Dark regions on the gravity image generally represent sedimentary basins, the dark signature being caused by a thick accumulation of low-density sediments. The 'blocky' character of the synergistic image (Plate 5.12) in the northeast is very evident and coincides with the Tertiary basalt outcrop. The different colour variations associated with the basalts are due to variations in the types of rock that lie beneath them. Northeast of the Tow Valley Fault (1, Figure 5.15) the brown/pink signature is due to the presence

of thick (c. 2 km) Permo-Triassic sediments. South of an ENE–WSW-trending lineament (2, Figure 5.15), the brown/orange signature again suggests the presence of sediments beneath the basalts. If this lineament is extrapolated westwards, it connects with the Tempo-Sixmilecross Fault. The yellow signature coincides with Carboniferous and Triassic sediments (3, Figure 5.15).

The green signature south of the Tow Valley Fault (4, Figure 5.15) is due to the presence of Dalradian metasediments beneath the basalts. To the northeast where Dalradian rocks are exposed (5, Figure 5.15), a green/yellow signature is observed which contrasts with the yellow–very bright yellow to the south where Devonian, Triassic and Cretaceous sediments occur (6, Figure 5.15). The prominent pink colour is the Lough Neagh depocentre (7, Figure 5.15), in which the separate ENE–WSW-trending fault-bounded troughs can be seen. The Sixmilewater Fault (8, Figure 5.15) forms the southern boundary to the northern basin.

The southeast corner of the image is dominated by a number of colours. An irregular red signature (9, Figure 5.15) is due to a Caledonian granite which is possibly at its thickest in the pink area (10, Figure 5.15). The shape and extent of the red signature is unlike the surface outcrop on published geological maps, suggesting that the granite extends at depth both northwards and westwards (11, Figure 5.15). A small dioritic intrusion east of the granite is associated with a conspicuous yellow colour (12, Figure 5.15). Diorite may also occur at depth to the southwest of the outcrop, where a similar signature can be discerned (13, Figure 5.15). The granite is intruded into Lower Palaeozoic greywackes which are associated with a dark grey/purple colour (14, Figure 5.15). The red associated with the granite contrasts well with the distinctive blue colour for the Tertiary Slieve Gullion gabbro to the south (15, Figure 5.15). The signature for the gabbro is identical to a 20 km diameter circular feature to the west (16, Figure 5.15) on which is superimposed a number of WNW–ESE-trending dykes. These Tertiary dykes are up to 40 km in length and 100 m wide and are associated with negative magnetic anomalies

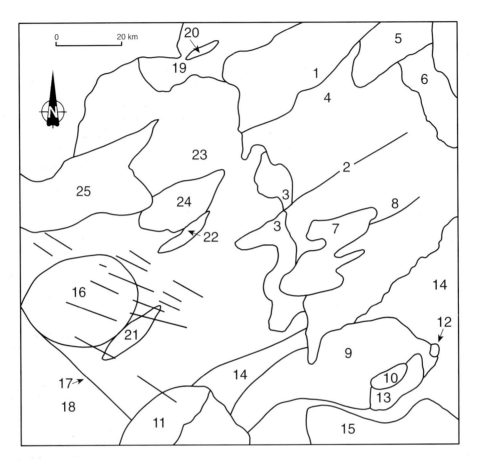

Figure 5.15 Interpretation of Plate 5.12. See text for details.

(Gibson and Lyle 1993), so that they tend to appear dark (17, Figure 5.15). The coincidence of a signature similar to the Slieve Gullion gabbro in combination with a major dyke swarm suggests the presence of a mafic body at depth which has been the magma source for the dykes. Carboniferous limestones and clastics crop out in this region and are associated with a bright yellow/orange signature (18, Figure 5.15). In the north (19, Figure 5.15) a similar signature is also observed for Carboniferous and Triassic sediments into which a dolerite (20, Figure 5.15) has been intruded. An elongated very dark blue colour (21, Figure 5.15) coincides with Carboniferous and Devonian sediments. However, the signature – which lies along part of the Tempo-

Sixmilecross and Clogher Valley Faults – is probably not related to these lithologies. A small inlier of Silurian greywackes is within this area and the colour probably reflects the presence of older rocks close to the surface. The Ordovician volcanics in Tyrone are shown by a blue signature (22, Figure 5.15). Colour variations in the west vary from blue (23, Figure 5.15) through green (24, Fig. 5.15) to yellow (25, Figure 5.15) and reflect lithological variations in the Dalradian, though direct lithological correlations are difficult. The Lough Foyle Succession tends to be blue whereas the Newtownstewart Quartzite group is yellow. The green areas approximate to the Mullaghcarn and Glenelly schists.

Conclusions

Digitally processed geophysical data can allow the gross geological features of an area to be mapped. In the case study presented here, ratioing and a principal components transform, similar to that applied to conventional satellite data, provided the optimum enhancement. Other enhancements, such as filtering of geophysical data, can be used to delineate faults (see Gibson 1993).

5.7 CHAPTER SUMMARY

- The case studies presented in this chapter illustrate some of the diverse applications for which remote sensing data can be used when suitably processed and enhanced.
- Low spatial, high temporal resolution imagery (Meteosat) is ideal for climatological studies such as rainfall assessment over West Africa. In some instances, such as tropical cyclone monitoring over the Arabian Sea, remote sensing may provide the only viable methodology for studying such storms.
- Higher spatial resolution sensors such as Landsat TM, combined with appropriate digital image processing, can prove useful for the investigation of land-cover variations in which individual land classes occupy small areas. Thus band ratioing and band decorrelation techniques were used in order to differentiate peat areas in northwest England.
- The versatility of current image processing software packages allows non-image data to be combined with conventional remotely sensed data in order to investigate specific problems. An example is provided in section 5.3 in which Landsat TM data were combined with digital data in order to model various sea level rise scenarios.
- The geological case study showed how digitally processed non-image datasets could be used to provide information that cannot be obtained from satellites, such as SPOT, which yield reflectance data.

FURTHER READING

Rainfall Monitoring in West Africa References
Barrett, E. C. (1974) *Climatology from Satellites*, London: Methuen.
Bunting, J. T. and Hardy, K. R. (1984) in A. Henderson-Sellers (ed.) *Satellite Sensing of a Cloudy Atmosphere*, 'Cloud identification and characterisation' in A. Henderson-Sellers (ed.) *Satellite Sensing of a Cloudy Atmosphere*, London: Taylor and Francis, 203–40.
Dennett, M. D., Elston, J. and Rodgers, J. A. (1985) 'A reappraisal of rainfall trends in the Sahel', *Journal of Climatology* 5: 353–61.
ESOC (1985) Meteosat-2 Calibration Report (Issue 13) for July–September 1985. MEP/MET, Meteosat Exploitation Project, ESOC Darmstadt, West Germany, October 1985.
Griffiths, J. F. (1972) *Climates of Africa*, World Survey of Climatology vol. 10, ed. H. E. Landsberg, Amsterdam: Elsevier, 193–221.
Leroux, M. (1983) *The Climate of Tropical Africa*, Paris: Champion.
Ojo, O. (1977) *The Climate of West Africa*, London: Heinemann.

Topographic Mapping and Sea-Level Rise Modelling References
Fox, D., Hindley, D. and Power, C. (1991) 'The determination of areas of peat in satellite imagery', NRSC Ltd, report DRA SP(91) WP1 06.
NCC Scotland (1987) *The Ecology and Management of Upland Habitats: the role of remote sensing*, Special Publication no. 2. Report on one-day workshop held at NCC Scotland HQ 24 July 1987.
Sader, S. A., Powell, G. V. N. and Rappole, J. H. (1991) 'Migratory bird habitat monitoring through remote sensing', *International Journal of Remote Sensing* 12, 3: 363–72.

Tropical Cyclone Tracking References
NOAA (1989) Storm Data and Unusual Weather Phenomena, vol. 31 no. 9, NOAA National Weather Service, September 1989.
Power, C. H. (1989) 'Use of satellite remote sensing of cloud and rainfall for selected operational applications in the fields of applied hydrology and food production', Ph.D. thesis, University of Bristol.
Power, C. H., Barrett, E. C. and Beaumont, M. J. (1989) 'A satellite climatology of Arabian Sea tropical storms and cyclones 1966–1987', Final report to the Sultanate of Oman Ministry of Environment and Water Resources contract no. C1/88.
WMO (1977) 'The use of satellite imagery in tropical cyclone analysis', WMO Technical Notes.

Peat Mapping References
Carroll, D. M., Evans, R. and Bendelow, V. C. (1977) *Air Photo Interpretation for Soil Mapping*, Soil Survey Technical Monograph 8, Harpenden.
Horgan, G. W. and Critchley, M. F. (1989) 'The use of Thematic Mapper Imagery for peatland mapping and mineral exploration in Ireland', in *Proceedings of a Workshop on Earthnet Pilot Project on Landsat Thematic Mapper Applications* held at Frascati, Italy, in December 1987, 91–7.

Knight, D. (1991) 'Growing threats to peat', *New Scientist* 131 (1780): 27–32.

Lynn, D. W. (1984) 'Surface material mapping in the English Fenlands using airborne multispectral scanner data', *International Journal of Remote Sensing* 5, 4: 699–713.

Middleton, R. (ed.) (1990) 'The North West Wetlands Survey Annual Report', Lancaster University Archaeology Unit.

Stove, G. C. and Hulme, P. D. (1980) 'Peat resource mapping in Lewis using remote sensing techniques and automated cartography', *International Journal of Remote Sensing* 1, 4: 319–44.

Geological Mapping References

Drury, S. A. and Walker, A. D. (1987) 'Display and enhancement of gridded aeromagnetic data of the Solway Basin', *International Journal of Remote Sensing* 8, 10: 1433–44.

Gibson, P. J. (1991) 'An integrated investigation of the Tow Valley Fault system, Ireland, with particular reference to remote sensing techniques', Ph.D. thesis, University of Ulster.

Gibson, P. J. (1993) 'Evaluation of digitally processed geophysical data sets for the analysis of geological features in northern Ireland', *International Journal of Remote Sensing* 14: 161–70.

Gibson, P. J. and Lyle, P. (1993) 'Evidence for a major Tertiary dyke swarm in County Fermanagh, Northern Ireland, on digitally processed aeromagnetic imagery', *Journal of the Geological Society* London 150: 37–8.

Lee, M. K., Pharaoh, T. C. and Soper, N. J. (1990) 'Structural trends in central Britain from images of gravity and aeromagnetic fields', *Journal of the Geological Society*, London 147: 241–58.

APPENDIX A

ANSWERS TO SELF-ASSESSMENT TESTS

CHAPTER 1

1 Polar-orbiting satellites with spatial resolutions less than 100 m include Landsat, SPOT, JERS-1 and IRS. The NOAA satellites have a much coarser spatial resolution.

2 The 8–14 µm range is within an atmospheric window. Measurements can thus be obtained by remote sensing sensors within this waveband.

3 Both water vapour and carbon dioxide absorb electromagnetic radiation at 2.7 µm.

4 Landsat differs from SPOT in a number of ways:
  (a) Landsat uses an across-track scanning system to obtain data whereas SPOT uses a pushbroom system.
  (b) SPOT has off-nadir viewing capabilities, unlike Landsat.
  (c) Stereo-images can be formed by using SPOT data.
  (d) Landsat obtains seven bands of data whereas SPOT can either obtain one band (panchromatic mode) or three bands (multispectral mode). SPOT 4 can obtain four bands of multispectral data.
  (e) Landsat TM and MSS have spatial resolutions of 30 m and 80 m respectively whereas SPOT has a spatial resolution of 20 m in multispectral mode and 10 m in panchromatic mode.

5 Important target parameters are the dielectric constant, slope, aspect and surface roughness and important system parameters are polarisation and wavelength.

6 A 10-bit system measures 1,024 grey levels (2^{10}) and the highest DN is 1,023.

7 Often two surfaces can only be differentiated at wavelengths which cannot be detected by the human visual system. Thus, for example, an infrared image is displayed in red and variations in red are consequently indicative of variations in the infrared signature.

CHAPTER 2

1 A skip factor of 4 means that every fourth pixel in every fourth line is sampled, i.e. one out of every 16. Therefore 6.25 per cent of the dataset is sampled.

2 A line dropout is a line within an image dataset for which no data have been obtained. Line banding is a defect in which every nth line is darker (or paler) than the rest of the scene. The data in such lines are real but they do not match the rest of the image. Data in a line dropout are lost completely.

3 A good ground control point must be easily identifiable on the unrectified image and on the georeferenced image or map. It must maintain a fixed position in space and ideally should not vary with time. Thus, for example, the edge of a sand bar would make a poor ground control point as it will migrate and change its position with time.

4 The Sun has a lower illumination angle (as measured from the horizontal) during winter in the northern hemisphere (see Figure 2.17a).

5 Polynomials are sets of equations that link the locations of the ground-control points in an unrectified image to a georeferenced map or image. They are subsequently used to rectify the entire unrectified image.

CHAPTER 3

1 From equation 3.1:

$$68 = 255 \times \frac{75 - 55}{x - 55}$$

Therefore $68 (x - 55) = 255 \times 20$

$$x - 55 = (255 \times 20)/68$$

$$= 75$$

Therefore $x = 75 + 55$

$$= 130.$$

2 The number of simple ratios is 28. This can be determined by creating the triangle below.

1/2   1/3   1/4   1/5   1/6   1/7   1/8

2/3   2/4   2/5   2/6   2/7   2/8

3/4   3/5   3/6   3/7   3/8

4/5   4/6   4/7   4/8

5/6   5/7   5/8

6/7   6/8

7/8

The value of 28 may also be determined by using equation 3.3 where n is 8 and r is 2. To calculate the number of potential false colour composites, you will require equation 3.3. In this instance n is 28 and r is 3, therefore the number of potential false colour composites is 3,276.

3 This question can be solved using equations 3.1 and 3.2. From equation 3.1, when stretched, 60 will have a value of:

$$255 \times \frac{60 - 20}{150 - 20}$$

$$= 78$$

and 45 has a value of:

$$255 \times \frac{45 - 20}{150 - 20}$$

$$= 49$$

Therefore the input range of 45–60 is increased to 49–78.

When the stretch is performed from 20 to 80, then from equation 3.2, 60 will have a value of:

$$255 \times \frac{60 - 20}{80 - 20}$$

$$= 170$$

and 45 has a value of:

$$255 \times \frac{45 - 20}{80 - 20}$$

$$= 106$$

Therefore the input range of 45–60 is increased to 106–70.

4 An edge is an abrupt change in the DN values over a short distance. Edges are important in geology because they often represent faults.

5 For Figure 3.33a, the input range 20–40 is linearly stretched, the range 41–70 is compressed in the output and the 71–90 input is again stretched. Values greater than 90 are saturated to 255. The LUT represented by Figure 3.33b shows compression of the input range 20–40 and a non-linear stretch from 41 to 80 followed by a linear stretch from 81 to 90. Values greater than 90 are saturated to 255.

6 The results of applying the filters are shown in Figure A1. The DN variation across the low-pass filtered image is less than the variation across the

original image, and thus the edge will be less prominent. Low-pass filters average out the DNs in an image and minimise abrupt changes. Taking the sixth line of data as an example, the DNs go from 44 to 2 to 44 across the edge (a change of 42 in both instances). In the low-pass image the change is from 29 to 26 to 23 (a change 3). However, on the high-pass filter, the edge is characterised by a larger DN change than the original, and thus it will appear sharper. Again for the sixth line, the change is from 59 to −22 to 65, a difference of 81 and 87.

CHAPTER 4

1 Six criteria which are essential for cloud identification are: brightness, texture, shape, size, organisation and shadow effects.

2 Rain-clouds can be differentiated from other clouds on visible imagery, as rain-clouds are bright, i.e. they have high grey-scale values, and on thermal infrared imagery they are cold, i.e. usually with high grey-scale values being equivalent to the coldest temperatures.

3 Satellite imagery can improve the assessment of rainfall and cloud cover in the following ways: First, in remote regions where rain-gauge and ground radar networks are sparse or non-existent. Satellite images often provide the only regular reliable record of rain-cloud and weather situations producing rainfall. Second, satellite images give a spatial coverage of rain-cloud areas from which rainfall amounts can be estimated, often filling in the gaps between rain-gauge and ground radar point location readings. Third, satellite images can be used to check whether there was any rain-cloud during the period when rainfall was recorded or observed by conventional means, or provide objective estimations of rainfall when conventional instruments are not operating properly.

4 Low-resolution imagery is better suited to weather system analysis as weather systems often cover thousands of kilometres and several countries. For their extent and influence to be properly appreciated, they must be observed as complete entities on satellite imagery which covers several countries in one image. As weather systems are so large and are constantly changing, they do not have to be observed in great spatial detail but do need to be observed frequently throughout a 24-hour period. Therefore it is better to receive several images a day from a low-detail high-altitude orbiting satellite that show the dynamic changes in a weather system than to receive a detailed image maybe only once or twice a month from a satellite in a close Earth orbit.

5 (a) A manual rainfall estimation technique is performed on hardcopy analogue satellite imagery by skilled satellite weather analysts.

   (b) An interactive technique combines both manual analyses by skilled weather analysts (using, for example, on-screen digitising or interactive data entry) with pre-programmed computer analysis, to produce final calculation analysis maps or assessments.

   (c) An automatic technique uses computer software to perform the complete analysis of rainfall from satellite imagery, the only human intervention being at the data input stage and the receipt of the output of results.

6 Healthy vegetation has its highest response in the near-infrared part of the spectrum between 0.7 and 0.9 μm, and relatively much lower responses in all other parts of the visible and infrared parts of the spectrum except the thermal infrared. No other Earth surface has this response.

7 A ratio of the near-infrared and red bands.

8 All vegetation indices are based on two assumptions: (1) that an algebraic combination of spectral bands can provide useful information about vegetation and (2) that all bare soil in an image will form a line, the soil brightness line, in spectral space, which is the line for zero vegetation.

9 This is a monthly calendar through the year giving the growth stages of a number of different crops for each month of the year, i.e. the seasonal cycle of ploughing, planting, emergence, growth, maturity, harvest and fallow for each crop. Each global region has its own crop calendar determined by local climate and farming practices. In the agricultural context, the visual representation of phenology shown on satellite imagery illustrates the local crop calendar. The effects of crop husbandry activities can be seen on satellite images. Crop identification, classification and interpretation are often improved by a knowledge of growth cycles of individual crops and the ability to compare winter with spring crops. Therefore by using several images obtained at different times throughout the season, the analyst can discriminate between different crops.

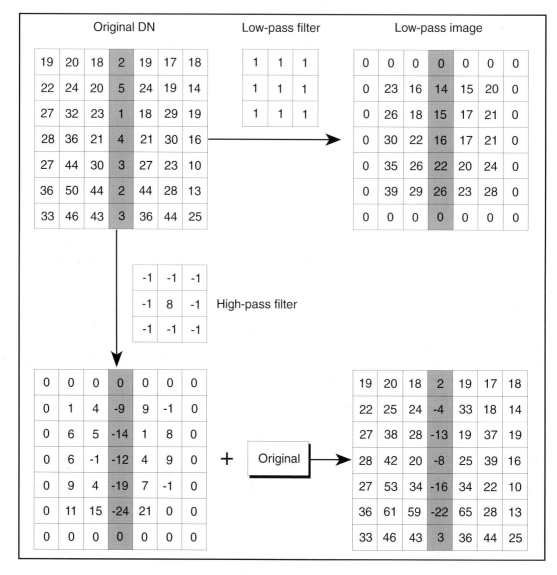

Figure A.1 Answer to question 6 in self-assessment test for Chapter 3.

APPENDIX B

ACRONYMS USED IN REMOTE SENSING

A bewildering array of acronyms exists in remote sensing and one could probably write a book on these alone. The following list indicates the main ones that have been mentioned in the text or which may be encountered in other remote sensing literature.

ADEOS	Advanced Earth Observing Satellite
AMI	Active Microwave Instrument
AMPS	Airborne Multisensor Pod System
ARTEMIS	African Real Time Environment Data and Information Service
ATSR	Along-Track Scanning Radiometer
AVHRR	Advanced Very High Resolution Radiometer
BIAS	Bristol-NOAA InterActive rainfall monitoring Scheme
BNSC	British National Space Centre
BURS	Bradford University Remote Sensing Ltd
CCD	Charge-coupled device
CCRS	Canadian Centre for Remote Sensing
CCT	Computer Compatible Tape
CD-ROM	Compact Disk Read Only Memory
CEC	Commission of European Countries/Communities
CEO	Centre for Earth Observation
CODE	Coastal Ocean Dynamics Experiment
CZCS	Coastal Zone Colour Scanner
DERA	Defense Evaluation and Research Agency
DMSP	Defense Meteorological Satellite Program
EOSAT	Earth Observation Satellite Company
EROS	Earth Resources Observation System

ERS	European Remote Sensing Satellite
ESA	European Space Agency
EUMETSAT	European Meteorological Satellite operations centre
FCC	False colour composite
FRONTIERS	Forecasting Rain Optimised using New Techniques of Interactively Enhanced Radar and Satellite data (former UK Meteorological Office weather forecasting system)
GAC	Global Area Coverage
GARP	Global Atmospheric Research Programme
GATE	GARP Atlantic Tropical Experiment
GCOS	Global Climate Observing System
GIS	Geographical Information System
GOES	Geostationary Operational Environmental Satellite
GOME	Global Ozone Monitoring Experiment
GMS	Geostationary Meteorological Satellite
GPS	Global Positioning System
HCMM	Heat Capacity Mapping Mission
HIRS	High Resolution Infrared Radiation Sounder
ICSU	International Council of Scientific Unions
IFOV	Instantaneous Field Of View
IJRS	International Journal of Remote Sensing
IOC	Intergovernmental Oceanographic Commission
IR	Infrared
IRS	Indian Remote Sensing satellite

JERS	Japanese Environmental Remote Sensing satellite	SeaWiFS	Sea-viewing Wide Field-of-view Sensor
JPL	Jet Propulsion Laboratory	SIR	Shuttle Imaging Radar
JSFC	Johnston Space Flight Centre	SMS	Synchronous Meteorological Satellite
LAC	Local Area Coverage	SPOT	Système Pour l'Observation de la Terre
LFC	Large Format Camera		
MEIS	Multispectral Electro-optical Imaging Scanner	SSM/I	Special Sensor Microwave Imager
		SSM/T-1	Special Sensor Microwave Tempera-ture Sounder
MIR	Mid Infrared		
MOS	Marine Observation Satellite	SSM/T-2	Special Sensor Microwave Water Vapour Sounder
MSS	Multispectral Scanner		
MSSL	Mullard Space Science Laboratory of University College London	SST	Sea Surface Temperature
		TDRS	Tracking and Data Relay Satellite
MTF	Modulation Transfer Function	TIMS	Thermal Infrared Multispectral Scanner
MTPE	Mission To Planet Earth		
NASA	National Aeronautics and Space Administration	TIR	Thermal Infrared
		TIROS	Television Infrared Operational Satellite
NASDA	National Space Development Agency		
		TM	Thematic Mapper
NCDC	National Climate Data Centre	TOMS	Total Ozone Mapping System
NDVI	Normalised Difference Vegetation Index	TOVS	TIROS Operational Vertical Sounder
		TRMM	Tropical Rainfall Mapping Mission
NOAA	National Oceanic and Atmospheric Administration	UARS	Upper Atmosphere Research Satellite
		UNEP	United Nations Environmental Pro-ject, based in Nairobi, Kenya
NRI	Natural Resources Institute of the University of Greenwich, UK		
		UNESCO	United Nations Educational, Scien-tific and Cultural Organisation
NRSC	National Remote Sensing Centre		
NSIDC	National Snow and Ice Data Centre	UN FAO	United Nations Food and Agricul-tural Organisation
OCTS	Ocean Colour and Temperature Scanner		
		UV	Ultraviolet
OLS	Operational Linescan System	VI	Vegetation Index
PDUS	Primary Data User Station	VIS	VISible waveband/s of satellite imagery
RADAR	RADio And Ranging		
RAR	Real Aperture Radar	VISSR	Visible Infrared Spin-Scan Radiometer
RBV	Return Beam Vidicon	WCRP	World Climate Research Programme
RGB	Red Green Blue colour composite image	WMO	World Meteorological Organisation
		WV	Water Vapour channel from GMS
SAR	Synthetic Aperture Radar	WWW	World Wide Web (Internet)

APPENDIX C

GLOSSARY OF REMOTE SENSING TERMS

This glossary provides a brief explanation of the main terms used in remote sensing.

Absorption band: The wavelengths of electromagnetic radiation that are absorbed by a substance. Some components of the atmosphere have a number of absorption bands. Water vapour in the atmosphere, for example, absorbs radiation with a wavelength of 6 μm.

Across-track scanning system: See transverse scanning system.

Active microwave instrument: Radar systems carried aboard the European Radar Satellites comprising a SAR and wind scatterometer.

Active remote sensing system: A remote sensing system that provides its own source of electromagnetic radiation to illuminate the target. Radar is the commonest active system employed in remote sensing. See also passive remote sensing.

Additive primary colours: The colours red, green and blue which, when added in different combinations, can produce all the colours.

Advanced Very High Resolution Radiometer: See AVHRR.

Aerial photograph: A representation of a scene recorded on photographic film from the air. Acquisition platform is usually an aircraft. See also oblique aerial photograph and vertical aerial photograph.

Aggregation: The merging of two or more separate classes into a single class. This may be used when there is a substantial overlap between classes.

Albedo: The albedo of a surface is the ratio of the radiation reflected from it to the total electromagnetic radiation incident on it.

Algorithm: The term applied to a series of instructions given to a computer.

ALMAZ: A former Russian radar satellite, mission life 1991–1993.

Along-Track Scanning Radiometer: A scanning system carried aboard the European Radar Satellites. On ERS-2, it comprises a seven-channel radiometer and a microwave sounder.

Along-track scanning system: See pushbroom system.

Altimetry: Technique for the measurement of the height of a satellite above the surface. The length of time a pulse of electromagnetic radiation takes on a round trip from the satellite to the ground and back is measured and from this the height can be accurately determined. When employed over oceans, the sea height can be measured.

Amplitude: Characteristic of an electromagnetic wave. It is the distance from peak to trough.

Atmospheric window: Atmospheric windows represent the wavelengths of electromagnetic radiation that are not absorbed by components of the atmosphere. The waveband 0.4–0.7 μm (visible) is an atmospheric window. Remote sensing systems are generally designed to operate within atmospheric windows.

AVHRR: Advanced Very High Resolution Radiometer. Multispectral scanner system aboard the TIROS satellites.

AVIRIS: Acronym for Airborne Visible Infrared Imaging Spectrometer. This is a remote sensing system which typically obtains data in many bands (> 200) with very high spectral resolution.

Azimuth direction: A term used in radar to describe the direction in which the airborne sensor is moving. See also range direction.

Azimuth resolution: A term used in radar to describe the resolution in the direction in which the sensor platform is moving. See also range resolution and synthetic aperture radar.

Backscatter: A term used in radar to describe the energy that is scattered back to the airborne or spaceborne platform.

Band: A wavelength range measured by a remote sensing system. Often used to designate how many 'layers' of data are recorded by a system, e.g. Landsat TM is a seven-band system. Note that the term 'channel' is occasionally used instead of band.

Band interleaved by line: A term that describes the format in which the digital data on a computer-compatible tape are arranged. In band interleaved by line format (BIL) for an n-band system, the first n lines of data hold the digital numbers for the first line for all n bands.

Band interleaved by pixel: A term that describes the format in which the digital data for a scene are arranged. In this format, for an n-band system, the first n numbers are the DNs for the first pixel on the first line, the second set of n numbers are the n digital numbers for pixel 2 line 1 and so on.

Band ratioing: An enhancement procedure in which a new image is created by dividing the DN in one band by the DN in another band for every pixel in the dataset. Band ratioing has particular applications in mineral exploration and in vegetation surveys. Also see vegetation index.

Binary number: Numerical system based on the number 2.

Bit: Abbreviation for binary digit; can be represented by zero or one. The term is analogous to quantum and represents the smallest discrete package of information. Landsat TM, for example, transmits its data at 85 megabits per second.

Blackbody: A theoretical body which absorbs the entire radiation incident on it and which obeys the Stefan–Boltzmann Law.

Brightness value: A term synonymous with digital number.

Byte: A collection of 8 bits (256 levels).

C: Letter assigned to the speed of light (and all electromagnetic radiation) which is equal to 3×10^8 m/s in a vacuum.

Camera: A device for recording the reflectance characteristics of a scene on a sensitised medium which generally operates within the visible or photographic infrared sections of the electromagnetic spectrum.

CCD: Charge-coupled device which is used as a detector on pushbroom systems.

CCT: Computer-compatible tape on which satellite data are stored.

CD-ROM: Compact Disk, Read Only Memory. An optical storage device which is commonly used to hold image data because such devices can store up to 600 Mb of data.

Central Processing Unit (CPU): Part of the computer system that processes the data according to the operator's instructions.

Classification: The procedure by which an n-band dataset is converted into a thematic image with a finite number of classes. See also unsupervised classification and supervised classification.

Coastal Zone Colour Scanner: A scanning system primarily designed to measure chlorophyll variations in the oceans which was carried on board Nimbus 7.

Complementary colour: The complementary colour of a primary colour is produced by mixing the two remaining primary colours. Thus yellow is the complement of blue because it is produced by green and red light.

Cone: Detectors in the human eye that are sensitive to colour. See also rod.

Contrast stretch: The procedure by which the input DN range for an image, which usually spans only a small part of the available DN, is increased to encompass a larger part of the available DN. This mathematical procedure increases the contrast of the image.

Co-registration: The alignment of different bands of data or data obtained from different systems so that the row and column co-ordinates for all pixels for all bands are identical, where each pixel represents a discrete ground area.

Corner reflector: A term used in radar to describe intersecting surfaces which return the incident energy back towards the sensor and thus produce a bright signature.

Correlation: In remote sensing, it is a measure of the degree to which the DN for one band can be predicted if the DN of another band is known.

Covariance: The joint variation of two variables about their common mean. The covariance is an important statistic used in some enhancement procedures such as principal components analysis.

Cross-polarised: A term applicable to a radar system which transmits in the vertical (or horizontal) direction and measures the returned signal in the horizontal (or vertical) plane. See also parallel-polarised.

CZCS: See Coastal Zone Colour Scanner.

DARTCOM: A UK company that has developed and markets a PC-based low spatial resolution satellite data reception system.

DD5: Russian high-resolution space photography system.

Depression angle: A term used in radar to describe the angle between a horizontal plane and the direct line from transmitter to the object that is being imaged. It decreases from near to far range.

Dielectric constant: A term used in radar that is a measure of the electrical properties of a substance.

Digital elevation model: A regular grid array of numbers in which the numbers represent elevation.

Digital image: An image obtained by a scanner system in which the measured parameter (such as reflectance) is represented by an array of pixels, each of which is associated with a digital number.

Digital image processing: The manipulation of digital data by computer programs in order to improve the appearance of an image.

Digital number: See DN.

Display monitor: Component of a digital image processing system on which the image is shown.

DN: Digital number. The number assigned to a pixel which is related to the parameter being measured by a remote sensing system. For example, a low DN may represent a low reflectance in a particular waveband and a high DN a high reflectance. DN values must be integers and common ranges used are 0–63, 0–128 and 0–255. Note: for an n-band dataset each pixel is associated with n digital numbers.

DN histogram: See frequency histogram.

Doppler shift: The apparent change in frequency as a wave moves past an observer. SAR systems employ this effect when recording data in order to produce a high azimuth resolution.

Dwell time: For a pushbroom system the length of time for which a ground-resolution cell is viewed by a detector. In a transverse scanning system, it is the time taken for the ground resolution cell to be swept by a scanning mirror.

Edge: An abrupt change in tone observed on an image.

Eigenvalue: A term used in principal components analysis that is a measure of the amount of variance in each principal component.

Eigenvector: A parameter that gives the orientation of a principal component axis.

Electromagnetic radiation: Energy transmitted at the speed of light by oscillating electric waves.

Electromagnetic spectrum: The range of wavelengths or frequencies of electromagnetic radiation.

Emissivity (ε): The ratio of the energy actually emitted by unit area of a surface in unit time at a given temperature to the energy emitted by unit area of a blackbody in unit time at the same temperature.

Emittance: See exitance.

EOSAT: US satellite data supplier and operating company.

ERDAS: US company that develops and sells image processing software.

ERMAPPER: Image processing software package developed by an Australian company, Earth Resources Mapping.

ESRIN: European satellite data collection and archiving centre.

EURIMAGE: European satellite data distribution centre/agency.

Exitance: Radiant flux of energy emitted by a body. Usually stated per unit area. Can be determined from the Stefan–Boltzmann Law.

False colour composite: An image whose colours do not accord with those that would be seen with the human eye. False colour composites usually include at least one input band that is invisible to our eyes. Projecting the green range in blue, the red range in green and the infrared range in red forms a standard false colour composite.

False colour image: See false colour composite.

Far range: A term used in radar to describe that part of the image which is farthest from the transmitting antenna. See also near range and depression angle.

Feature–space plot: A graph showing the distribution of digital numbers in a dataset in which the axes are formed by different spectral bands.

Fiducial: Tick marks on the edges or corners of aerial photographs that can be used to determine the principal point of the photograph.

Filter: A mathematical procedure that alters the digital numbers in an image. See also high-pass and low-pass filter.

Foreshortening: A term employed in radar to describe the compression effect observed on slopes being illuminated by a radar beam. Also see layover.

Frequency: A property of an electromagnetic wave that is a measure of the number of wave crests passing a point in unit time. It is inversely proportional to wavelength.

Frequency histogram: A bar chart representation of the DN variation in an image. It is produced by plotting DN values (0–255) on the horizontal axis against frequency of occurrence (vertical axis). Each band will have a different frequency histogram.

Geographical Information System: A system incorporating a collection of spatially correlated datasets and a collection of computer programs with which to analyse the data.

Geometric rectification: The process by which a distorted image is corrected, usually in such a way that north is at the top of the image and the scale is constant throughout the image.

Geostationary orbit: An orbit with a period of one solar day. The satellite thus always appears to remain stationary in the sky and is at a height of 36,000 km above the Earth.

Global Ozone Monitoring Experiment: See GOME.

GOME: Global Ozone Monitoring Experiment. ERS-2 carries a spectrometer which takes measurements between 0.24 μm and 0.79 μm which allow the concentrations of atmospheric constituents such as ozone, nitrogen oxide and nitrogen dioxide to be determined.

Ground-collected verification data: A term employed in remote sensing to describe the validation of a signature obtained by a remote sensing system by an inspection of the imaged site. (Some authors refer to this as 'ground truth').

Ground-control point (GCP): A point that can be located on both an image and a map which is used in the process of rectification. Such points are used to produce the mapping algorithms which change a distorted image into a geometrically correct one.

High-pass filter: A filter that accentuates edges and sharpens up an image. See also low-pass filter.

Histogram equalisation stretch: A non-linear contrast stretch which stretches the data depending on the number of DNs in each DN bin.

HRV: Acronym for High Resolution Visible that refers to the two sensors carried by SPOT.

Hybrid ratio image: A colour image in which, for example, two input images may be ratio images while the third input is an image obtained directly by the satellite. Thus, if the ratio of TM 1 to TM 2 is projected in red, the ratio of TM 3 to TM 4 is projected in green and TM 5 is projected in blue, the result is a hybrid ratio image. Such images may be useful as they re-introduce topographic expression into the display.

IFOV: See Instantaneous Field Of View.

IHS: An alternative means of visualising a colour image which does not use the conventional red, green, blue axes. The image is considered in terms of its intensity, hue and saturation. It is possible to convert an RGB image into an IHS image.

Illumination angle: The angle that the radiation from the Sun makes with the Earth as measured from the horizontal. This varies throughout the year, being smaller in the winter compared with the summer for the northern hemisphere. There is a lower annual variation near the equator.

Illumination direction: The direction from which the illumination comes when an area is being imaged. Satellite systems such as Landsat obtain their data in the morning when the Sun's illumination is from the southeast in the northern hemisphere. The illumination direction for active systems depends on the transmitter's orientation.

Image: A general term used to describe the representation of a scene obtained either photographically or digitally.

Image enhancement: The process by which data displayed on an image can be made more obvious to the human visual system.

Incidence angle: The angle that electromagnetic radiation makes with the surface measured from the vertical. This term is often applied to radar systems.

Instantaneous field of view: The solid angle through which a detector onboard a remote sensing system obtains data.

Intensity Hue Saturation: See IHS.

Interferometry: The process by which a three-dimensional surface can be created by using phase differences in the returned signals for active remote sensing systems.

Kinetic temperature: Temperature of a body measured by a thermometer. See also radiant temperature.

KVR-1000: Russian space panchromatic camera with 2 m spatial resolution.

Lambertian reflector: A surface that is rough compared with the wavelength of the radiation incident on it. The incident energy is consequently scattered in all directions.

Landsat: A series of Earth-observing satellites, the first of which was launched by the United States in 1972.

Layover: A term used in radar to describe the effect of the tops of tall features being closer to the radar system than the base of the features. Also see foreshortening.

LIDAR: Acronym for Light Intensity Detection and Ranging. LIDAR remote sensing systems are active and employ lasers to map topographic variations accurately or to identify particular substances.

Line banding: A defect observed on some images in which a regular banding effect is observed where lines of data are consistently darker or lighter than the rest of the scene. It appears on multi-detector systems in which a detector's response no longer matches the other detectors onboard the platform. See also sixth-line banding.

Line dropout: The loss of a line of data. This defect can be caused by a loss of signal or an error in the recording. DN values are zero for such a line. An image with a number of line dropouts may be visually improved by suitable image processing.

Lineament: A linear, topographic, tonal or textural feature observed on an image.

Logarithmic stretch: A non-linear stretch that preferentially stretches the darker parts of a scene (lower digital numbers).

Look angle: A term used in radar that is equal to 90 degrees minus the depression angle.

Low-pass filter: A filter that accentuates the low-frequency component of an image, generally 'blurring' the image.

Map projection: Process by which the elliptical Earth is represented on a two-dimensional surface.

Maximum likelihood classification: Classification procedure based on the assumption that training area datasets have a normal distribution, which involves the construction of probability contours.

Mean: A statistical measure of the average of a collection of numbers. It may be obtained by adding the numerical values and dividing by the number of measurements.

Median: The numerical value that divides a DN distribution into two equal halves.

Meteosat: Geostationary satellite that obtains data for Africa and Europe.

Microwave: Part of the electromagnetic spectrum ranging from 0.1 cm to 1 m.

Minimum distance classification: Classification procedure by which an unknown pixel is assigned to the class to which it is nearest, i.e. shortest distance to the mean of the class.

MK-4: Russian high spatial resolution space camera.

Mode: A term used in statistics that refers to the numerical value which occurs most often in a range of numbers.

Mosaic: A number of aerial photographs or satellite images joined together in order to display a larger area.

MSS (Multispectral scanner): A system that obtains data in a number of wavebands simultaneously. The MSS system may be carried by an airborne or spaceborne platform. The Landsat multispectral system operates at four wavebands.

Multispectral imagery: Imagery of an area that has been recorded in a number of wavebands. Landsat MSS, for example, records four bands.

Multispectral scanner: See MSS.

Multitemporal imagery: Imagery of the same area that has been obtained on different dates. See also multispectral imagery.

Near range: A term used in radar to describe that part of the image which is nearest to the transmitting antenna. See also far range and depression angle.

NOAA: Acronym for National Oceanic and Atmospheric Administration. This organisation maintains a number of meteorological satellites.

Non-linear stretch: A contrast stretch that stretches DN ranges by different amounts. Also see histogram equalisation, logarithmic and power-law stretch.

Normalised Difference Vegetation Index (NDVI): A ratio image used in vegetation studies produced by subtracting the DNs for the red band from the infrared band and dividing the result by the addition of the red and the infrared i.e. $(IR - R)/(IR + R)$. See also vegetation index.

Oblique aerial photograph: An aerial photograph obtained by a tilted camera system.

Parallelepiped classification: Classification procedure in which an n-dimensional box encompasses the training area and unknown pixels that fall within the box are assigned to the training-area class.

Parallel-polarised: A term applicable to a radar system which measures the returned signal in the same plane as it transmits the microwave radiation. See also cross-polarised.

Passive remote sensing: Remote sensing which uses the radiation from the Sun as the source of illumination. See also active remote sensing.

Photographic infrared: The part of the electromagnetic spectrum from 0.7 μm to 0.9 μm.

Pixel: Picture element; the smallest definable unit on a digital image.

Power-law stretch: A non-linear stretch that preferentially stretches the brighter parts of a scene (higher digital numbers).

Principal components analysis: Mathematical transform that produces new images based on the variance of the multispectral dataset.

Principal point: Centre point of an aerial photograph.

Pseudocolour image: A single-band black and white image in which each digital number or range of digital numbers is assigned a colour.

Pushbroom system: Imaging system which employs an array of many small detectors, each of which measures the reflectance for an individual ground sampling cell. SPOT uses a pushbroom imaging system.

Radar: Acronym for RAdio Detection And Ranging. It is the principal form of active remote sensing and operates within the microwave part of the electromagnetic spectrum.

Radar range: See near and far range

Radar rough: A term used in radar to describe a surface which backscatters a substantial amount of the incident energy towards the antenna. A radar-rough surface will appear bright on an image. See also radar smooth.

Radar smooth: A term used in radar to describe a surface which backscatters very little of the incident energy towards the antenna. A radar-smooth surface will appear dark on an image.

Radiance: The power emitted by a body per unit area per unit steradian.

Radiant temperature: Temperature of a body measured by a radiometer. For a real body (as opposed to a theoretical blackbody) the radiant temperature is always lower than the kinetic temperature.

Radiometric resolution: The number of grey levels measured by a digital system. An 8-bit system measures 256 grey levels; the available digital numbers thus range from 0 to 255.

RAR: A Real Aperture Radar is a system which employs a long antenna in order to improve the azimuth resolution. See also SAR.

Ratio image: An image produced by dividing the DNs in one band by the corresponding DNs in another band. It is often used in vegetation studies.

Rayleigh scattering: See scattering.

Real Aperture Radar: See RAR.

Reflected infrared: The part of the electromagnetic spectrum ranging from 0.7 μm to 3.0 μm.

Remote sensing: The acquisition and recording of information about an object without being in direct contact with that object.

Resolution: See radiometric, spatial, spectral and temporal resolution.

Rod: Type of detector used by the human visual system which detects variations in brightness. See also cone.

Roughness: See radar roughness.

SAR: Acronym for Synthetic Aperture Radar. A SAR system employs the principles of the Doppler shift to improve resolution in the azimuth direction. See also RAR.

Satellite: In remote sensing, an unmanned spacecraft which orbits the Earth obtaining data.

Scale: The scale of an image relates the distance on the ground between two objects and the distance between the objects as measured on an image. It is generally expressed as a ratio or as a fraction.

Thus a scale of 1: 10,000 means that a distance of 1 mm on an image is equivalent to 10 m.

Scattering: The random propagation of electromagnetic radiation as a result of interaction with various components of the atmosphere. Selective scattering is wavelength dependent and, depending on the relative size of the particles compared with the wavelength of the radiation, either Rayleigh or Mie scattering may occur. Rayleigh scattering is inversely proportional to the fourth power of wavelength and Mie scattering is inversely proportional to wavelength. Non-selective scattering is not dependent upon wavelength.

Scatterplot: Two (or three)-dimensional plot showing the DN distribution between bands.

SEASAT: Radar satellite launched in 1978 that was primarily designed to obtain data on the oceans.

SeaWiFS: Sea-viewing Wide Field-of-view Sensor. Satellite-based remote sensing system designed to obtain data about phytoplankton variations in the oceans.

Shuttle Imaging Radar: Radar experiments performed onboard the Space Shuttle. To date SIR–A/B/C have been carried out.

Signal-to-noise ratio: The ratio of the signal from the ground which carries information to the noise due to aberrations in the electronics or defects in the scanning system which degrade the signal to some extent.

Signature: The various elements that, combined, describe how a particular feature appears on an image. For example, a rock type may be characterised by its colour and texture.

SIR: See Shuttle Imaging Radar.

Sixth-line banding: A defect observed on some Landsat MSS images in which every sixth line is darker or lighter than the surrounding lines. It is caused by the degradation of a detector but may be rectified by suitable processing.

Slant-range: A term used in radar, the direct distance from a radar source to the target.

Space Shuttle: Reusable United States space vehicle which has been used to launch satellites and on which remote sensing experiments have been performed.

Spatial resolution: A measure of the amount of detail that can be observed on an image.

Spectral resolution: A measure of the number and width of bands obtained by a remote sensing system.

Specular reflector: A surface that is smooth compared to the wavelength of the radiation that is hitting it, so that the incident energy is reflected at the same angle as that at which it strikes the surface. Also see Lambertian reflector.

SPOT: Series of French polar-orbiting satellites. To date, four have been launched.

Standard deviation: A statistical measure of the spread or dispersion of data about its mean. Many programs in digital image processing such as principal components analysis or classification calculate and use this parameter.

Standard false colour composite: See false colour composite.

Stereoscope: An optical device which allows the simultaneous examination of a pair of overlapping aerial photographs to produce a three-dimensional effect.

Sun-synchronous: A term applied to a satellite's orbit such that the plane of the satellite's orbit is always at the same angle to the Sun.

Supervised classification: A classification process in which an image is separated into a number of information classes based on the statistical characteristics of training areas outlined by the operator. See also unsupervised classification.

Surface scattering: A term used in radar to indicate that the main scattering occurs at the surface. See also volume scattering.

Synergistic display: A display in which images from different systems are co-registered and displayed simultaneously. For example, SPOT data may be combined with radar data. Such images may allow different surfaces to be discriminated more easily.

Synthetic Aperture Radar: See SAR.

Temporal resolution: A measure of how often an area is imaged by a particular remote sensing system. Landsat 5, for example, has a repeat cycle of 16 days.

Texture: A qualitative description of the rate of change of tone on an image.

Thematic map: A map produced by classifying an image in which a colour represents a specific theme such as land cover.

Thematic Mapper: This is a scanning system that has been carried by Landsats 4 and 5. It obtains data in seven wavebands with a 30 m resolution in bands 1–5 and 7 and 120 m resolution in band 6 (thermal).

Thermal capacity: A measure of the ability of a substance to retain heat.

Thermal conductivity: A measure of the ability of a substance to transfer heat.

Thermal inertia: A measure of how a body responds to changes in temperature.

TM: See Thematic Mapper.

Training area: An area on an image which the operator, using prior knowledge, knows belongs to one specific class which will be used in a supervised classification procedure. Thus, for example, an operator may delineate a known forest as a training area and use the generated signature for this area to search for other regions on the image which have the same DN characteristics as the training area and which may subsequently be assigned to the forest surface class.

Transverse scanner: A scanner system which employs an oscillating mirror which sweeps across the terrain in parallel strips which are at right angles to the platform movement.

Ultraviolet radiation: Part of the electromagnetic spectrum ranging from 0.03 μm to 0.4 μm.

Unsupervised classification: An automatic classification process in which the image is divided into a number of spectral classes based on DN distribution. The operator has minimal input. See also supervised classification.

US CROPCAST: Commercial product derived from meteorological satellite data giving updates on crop status/health throughout the growing season.

Variance: A statistical parameter used to measure the spread or the dispersion about the mean. The square root of variance is the standard deviation.

Vegetation index: A ratio image used in vegetation studies produced by dividing the data in the infrared band by the DNs in the red, i.e. (IR/R). See also Normalised Difference Vegetation Index.

Vertical aerial photograph: An aerial photograph taken with the camera pointing vertically down. Such photographs are obtained for stereoscopic viewing.

Visible light: That part of the electromagnetic spectrum that can be detected by the human visual system. It extends from 0.4 μm to 0.7 μm.

Volume scattering: A term applied to radar in which the radar signals are returned after interacting with the interior of a target such as a forest canopy. See also surface scattering.

Waveband: A term used to describe a range of the electromagnetic spectrum. Thus MSS 7 obtains data in the 0.7–1.1 μm waveband.

Wavelength: Characteristic of an electromagnetic wave, being the distance from peak to peak.

Whiskbroom scanner: See transverse scanner.

X-rays: Electromagnetic radiation with a very short wavelength (approximately 10^{-10} m). It is not used in remote sensing of the Earth.

APPENDIX D

REMOTE SENSING SOURCES OF INFORMATION

The general term 'sources of information' is used in the title of this appendix because it encompasses a range of themes. One might, for example, simply require information on the satellites that are currently in operation and the characteristics of their sensors. Alternatively, one may not require such information but need to know where specific datasets can be obtained or what software is available to process the data. Many remote sensing resources exist and it would not be possible to list them all. However, it is possible to categorise them broadly into government agencies, educational facilities and private companies. It should be realised that these categories are not mutually exclusive: university-based campus companies are today quite common. The nature of the 'service' provided also varies greatly. Some centres may act as repositories for digital data which may be purchased, whereas others may be contracted to undertake specific projects or even provide digital image processing facilities which may be hired for a short duration. The addresses of companies and institutions that have contributed imagery or data for volumes I and II of *Introductory Remote Sensing* are listed in Appendix E.

Many third-level institutions and universities have a remote sensing capability, even if it is quite elementary, and are usually willing to provide information on remote sensing to interested parties. Generally (though not exclusively!) remote sensing is not a separate subject but is housed in the Department of Geography, Department of Geology or the Department/School of Environmental Studies.

Various texts have been listed at the end of each chapter which readers should also consult if they wish to obtain a greater insight into remote sensing. A number of remote sensing atlases have been produced, an examination of which is useful for building up experience of the signatures of various features. Examples include:

- *Images of the World* (1984) published by Collins and Longman.
- *Earthwatch: a survey of the world from space* (1981) by C. Sheefield and published by Sidgewick and Jackson.
- *Man on Earth: the marks of man, a survey from space* (1983) by C. Sheefield and published by Sidgewick and Jackson.
- *NASA: views of Earth* (1985) by R. Kerrod and published by Admiral Books.
- *Britain from Space* (1985) by R. K. Bullard and R. W. Dixon-Gough and published by Taylor and Francis.
- *SEASAT Views North America, the Caribbean and Western Europe with Imaging Radar* (1980) by J. P. Ford, R. G. Blom, M. L. Bryan, M. I. Daily, T. H. Dixon, C. Elachi and E. C. Xenos, JPL Publication 80–67.
- *Space Shuttle Columbia Views the World with Imaging Radar: the SIR-A experiment* (1983) by J. P. Ford, J. B. Cimino and C. Elachi, JPL Publication 82–95.
- *Shuttle Imaging Radar Views the Earth from Challenger: the SIR-B experiment* (1986) by J. P. Ford, J. B. Cimino, B. Holt and M. R. Ruzek, JPL Publication 86–10.
- *Thematic Mapping from Satellite Imagery* (1988) by J. Denegre and published by Elsevier Applied Science Publishers Ltd.

- *The Home Planet* (1988) edited by K. W. Kelley and published by Queen Anne Press.

Although each chapter in this book details further readings, a difficulty with providing a 'further reading' section in a subject like remote sensing is that, because it is evolving quite rapidly, some of the references may soon be out of date. In order to be aware of the most current up-to-date research in aspects of remote sensing, it is necessary to consult the specialised academic journals which publish the results of such research. Notable journals include:

International Journal of Remote Sensing
Remote Sensing of the Environment
Canadian Journal of Remote Sensing
Institute of Electrical and Electronic Engineers: Transactions on Geoscience and Remote Sensing
Photogrammetric Engineering and Remote Sensing
Sistima Terra.

In addition, there are magazines which can also provide up-to-date information such as *SPOT* magazine by SPOT IMAGE, or *Earth Observation Quarterly*, which is published by ESA. As the reader is aware, remote sensing is employed in various fields; consequently, papers which involve the use of remote sensing appear in other journals. Thus, for example, the *Geological Journal* and the *Journal of the Geological Society, London*, have at different times published papers where remote sensing has been employed in a geological context. Meteorological journals such as *Weather, Atmospheric Environment* and the *International Journal of Climatology* regularly publish papers in which the interpretation of remote sensing data is an important component. Most geographical journals include occasional papers with a remote sensing theme and journals such as *Transactions of the Institute of British Geographers, Annals of Association of American Geographers* and *Progress in Physical Geography* should also be consulted.

Regular conferences on remote sensing take place, with an increasing trend towards specialisation. Thus one conference may concentrate on a specific theme such as applications of remote sensing for forestry while another may concentrate on the applications of a specific sensor, such as the ERS radar system. Information about these conferences may generally be obtained from third-level institutions or from societies such as the Remote Sensing Society (c/o Department of Geography, University of Nottingham, Nottingham, UK) or the American Society for Photogrammetry and Remote Sensing, 5410 Grosvenor Lane, Suite 210, Bethesda, MD 20814–2160.

Companies which supply digital-image processing software that can be used to analyse remote sensing data include:

EASI/PACE: IS Ltd, Atlas House, Atlas Business Centre, Simonsway, Manchester M22 5HF, UK, or PCI Enterprises, Richmond Hill Enterprises, Ontario L4B 1MS, Canada.

ENVI: Floating Point Systems UK Ltd, Ash Court, 23 Rose St, Wokingham, Berkshire RG40 1XS, UK.

Dimple: Cherwell Scientific Publishing Ltd, The Magdalen Centre, Oxford Science Park, Oxford OX 4GA, UK.

ERDAS Imagine: 2801 Bufford Highway, Suite 300, Atlanta, GA 30329–2137, USA, or Telford House, Fulbourn, Cambridge CB1 5HB, UK.

ERMAPPER: Earth Resource Mapping, 4370 La Jolla Village Drive, Suite 900, San Diego, CA 92122-1253, USA, or Blenheim House, Crabtree Office Village, Eversley Way, Egham, Surrey TW20 8RY, UK.

DRAGON: Goldin–Rudahl Systems Inc., 6 University Drive, Amherst, MA 01002, USA, or IS Ltd, Atlas House, Atlas Business Centre, Simonsway, Manchester M22 5HF, UK.

Idrisi: Idrisi Resources Centre, School of Earth and Environmental Sciences, University of Greenwich, Chatham Maritime, Kent ME4 4TB, or The Idrisi Project, Clark University, Laboratories for Cartographic Technology and Geographic Analysis, 950 Main Street, Worcester, MA 01610, USA.

IGIS: LaserScan Ltd, Cambridge Science Park, Milton Road, Cambridge, CB4 4FY, UK.

RSVGA: Eidetic Digital Imaging, Brentwood Bay, BC V0S 1AO, Canada.

V-image: VYSOR Integration Inc., Gatineau, Quebec J8T 5W5, Canada.

TeraVue: Geo-Services International (UK) Ltd, Unit 5, Des Roches Square, Witan Way, Witney, Oxfordshire OX8 6BE, UK.

TNTMips: Nigel Press Associates Ltd, Edenbridge, Kent TN8 6HS, UK.

One of the most extensive remote sensing resources is the World Wide Web (WWW), which consists of a networked system of information provided from a range of sources. A user in Paris can log on to data held by an organisation in New York. Search engines on the WWW allow the input of keywords in order to locate the relevant sites. However, a disciplined approach is needed to obtain specific information. The term 'remote sensing' will elicit over 300,000 references. Because of the extensive linkages between different sites on the WWW, there is no unique pathway to a specific piece of information. The following sites indicate some of the information that can be obtained from the WWW. The sites listed here (shown within square brackets) are correct at the time of writing (1998) though some may have changed address or ceased functioning since then.

The core of remote sensing is imagery and one may examine thousands of images from a number of sites. As regards radar imagery, the NASA/JPL Imaging Radar Site

[http: //southport.jpl.nasa.gov]

provides superb examples of data obtained during the SIR-C mission in 1994. Currently, the imagery is divided into a number of categories (e.g. archaeology, geology, glaciers and oceanography) and each image is accompanied by a description. A facility on this site allows one to download either the image or the data. Typically an image is 300 Kilobytes in size and in a jpeg or gif format, though the full datasets are substantially larger (15–60 Megabytes). However, the raw data can then be digitally processed to enhance different features. The range of features displayed on these datasets, and the amount of data available, make it possible to produce a large number of images designed to display an extensive range of environmental and cultural features which can form the basis of a teaching programme. NASA provides an image gallery site

[http: //www.nas.gov/gallery/photo/index.html]

which links to a range of other NASA sites such as the Johnston Space Flight Center, which is a repository for thousands of images obtained on manned space missions. Many of these images relate to the earliest United States missions (Mercury, Gemini and Apollo) though imagery obtained from the Skylab missions in the 1970s and Space Shuttle flights in the 1980s and 1990s are also included. Most of the images are photographs taken by the astronauts. Ocean colour data from the CZCS and SeaWiFS can be obtained at

[http: //daac.gsfc.nasa.gov/].

The Canadian Centre for Remote Sensing allows the user to observe different parts of Canada with different types of remote sensing sensors. MEIS, SPOT, TM and radar imagery is included. The site location is

[http: //www.ccrs.nrcan.gc.ca/ccrs/xhomepage.html].

It is possible to view a selection of NOAA images mostly of storms and hurricanes at a National Climate Data Centre site:

[http://www.ncdc.noaa.gov/ol/satellite/olimages.html].

The Sunday Times 'Windows on the World', CD-ROM, 1998.

APPENDIX E

CONTRIBUTORS TO *INTRODUCTORY REMOTE SENSING*

Texts such as *Introductory Remote Sensing: Principles and Concepts* and *Introductory Remote Sensing: Digital Image Processing and Applications* rely heavily on the good-will of many individuals and organisations to provide relevant remote sensing imagery and data and the permission to use them. We would like to thank the organisations listed below who made such contributions. It is invidious to pick out individuals who provided a great degree of assistance but especial thanks go Martin Critchley of ERA-Maptec, Hervé Lemeunier of SPOT IMAGE and Annie Richardson of JPL who provided most of the datasets included on the CD for this volume.

American Society for Photogrammetry and Remote Sensing
5410 Grosvenor Lane, Suite 210, Bethesda, MD 20814-2160, USA.

Battelle
Pacific Northwest Laboratories, Battelle Boulevard, PO Box 999, Richland, WA, 99352, USA.

Cambridge Committee for Aerial Photography (CUCAP)
Mond Building, Free School Lane, Cambridge CB2 3RF, UK.

Canada Centre for Remote Sensing
588 Booth Street, Ottawa, Ontario, K1A 0Y7, Canada.

CEN
100 Franklin Square Drive STE210, Somerset, NJ 08873, USA.

Department of Archaeology
National University of Ireland, Cork, Co. Cork, Republic of Ireland.

Department of Computer Science
National University of Ireland, Maynooth, Co. Kildare, Republic of Ireland.

Department of Geography
National University of Ireland, Maynooth, Co. Kildare, Republic of Ireland.

Discovery Programme
13 Hatch Street Lower, Dublin 2, Republic of Ireland.

ERA-Maptec
36 Dame St, Dublin 2, Republic of Ireland.

ESA/ESRIN
Via Galileo Galilei, I-00044 Frascati, Italy.

Goddard Spaceflight Centre
Greenbelt, MD 20771, USA.

Goldin–Rudahl Systems Inc.
6 University Drive, Amherst, MA 01002, USA.

Hunting Aerofilms Limited and Hunting Technical Services
Gate Studios, Station Road, Borehamwood, Hertfordshire WD6 1EJ, UK.

IS Limited
Atlas House, Atlas Business Centre, Simonsway, Manchester M22 5HF, UK.

Jet Propulsion Laboratory
California Institute of Technology, 4800 Oak Grove Drive, Pasadena, CA 91109-8099, USA

Johnston Spaceflight Center
Houston, TX 77058, USA.

National Aeronautics and Space Administration
Headquarters, Washington DC 20546-0001, USA.

National Oceanic and Atmospheric Administration
Satellite Data Services, World Weather Building, Washington DC 20230, USA.

National Remote Sensing Centre
Delta House, Southwood Crescent, Southwood, Farnborough, Hampshire GU14 0NL, UK.

National Snow and Ice Data Center
CIRES CB 449, University of Colorado, Boulder, CO 80309-0449, USA.

National Space Development Agency of Japan
2-4-1 Hamamatsu-Cho, Minato-Ku, Tokyo, Japan.

Natural Resources Institute
University of Greenwich, Medway Campus, Chatham Maritime, Kent ME4 4TB, UK.

Ordnance Survey of Ireland
Phoenix Park, Dublin 6, Republic of Ireland.

Ordnance Survey of Northern Ireland
Colby House, Stranmillis Court, Belfast BT9 5BJ, Northern Ireland.

RADARSAT International Inc.
Satellite Data Distribution Centre, 3851 Shell Road, Suite 200, Richmond, BC V6X 2W2, Canada.

Remote Sensing Unit
Department of Geography, University of Bristol, University Road, Bristol BS8 1SS, UK.
Now: Remote Sensing Centre, School of Geographical Sciences.

Sandia National Laboratories
PO Box 5800, Albuquerque, NM 87185-0529, USA.

School of Earth and Environmental Sciences
University of Greenwich, Medway Campus, Chatham Maritime, Kent ME4 4TB, UK.

Space Imaging
9361 Grant Street, Suite 500, Thornton, CO 80229, USA.

SPOT IMAGE
5 rue des Satellites, BP 4369, F-31030 Toulouse Cédex 4, France.
Tel. no. (33) 562194101
Fax no. (33) 562194054

Ten-to-Ten (DIGITECH) Ltd
Unit 6D, Aberystwyth Science Park, Aberystwyth SY23 3AH, Wales.

United States Geological Survey
EROS Data Center, Sioux Falls, SD 57198, USA.

APPENDIX F

FURTHER READING

Adler, R. F. and Negri, A. J. (1988) 'A satellite infrared technique to estimate tropical convective and stratiform rainfall', *Journal of Climate and Applied Meteorology* 27: 30–51.

Arino, O., Vermote, E. and Spaventa, V. (1998) 'Operational atmospheric correction of Landsat TM imagery', *Earth Observation Quarterly* 56: 32–5.

Arkin, P. A. (1979) 'The relationship between fractional coverage of high cloud and rainfall accumulations during GATE over the B-scale array', *Monthly Weather Review* 107: 1382–7.

Atkinson, P. H. and Tatnall, A. R. L. (1997) 'Neural networks in remote sensing', *International Journal of Remote Sensing* 18, 4: 699–709.

Atlas, D. and Thiele, D. W. (eds) (1981) *Precipitation measurements from space*, NASA Goddard Space Flight Center, Greenbelt, MD.

Austin, G. L. and Bellon, A. (1982) 'Very short range forecasting of precipitation by the objective extrapolation of radar and satellite data,' in K. A. Browning (ed.) *Nowcasting*, London: Academic Press: 177–90.

Avery, T. E. and Berlin, G. L. (1992) *Fundamentals of Remote Sensing and Airphoto Interpretation*, 5th edition, Englewood Cliffs, NJ: Prentice-Hall.

Ayres, F. (1983) *Theory and Problems of Matrices*, Singapore: McGraw Hill Book Co.

Bailey, J. O., Barrett, E. C. and Kidd, C. (1986) 'Satellite passive microwave imagery and its potential for rainfall estimation over land', *Journal of the British Interplanetary Society* 39: 527–34.

Ban, Y. and Howarth, P. J. (1996) 'Integration of ERS-1 SAR and Landsat TM data for agricultural crop classification' in *Proceedings of the 26th International Symposium on Remote Sensing of Environment*.

Ban, Y., Treitz, P. M., Howarth, P. J., Brisco, B. and Brown, R. J. (1995) 'Improving the accuracy of synthetic aperture radar analysis for agricultural crop analysis', *Canadian Journal of Remote Sensing* 21, 2: 158–64.

Baret, F. and Guyot, G. (1991) 'Potentials and limits of vegetation indices for LAI and APAR assessment', *Remote Sensing of the Environment* 35: 53–70.

Baret, F., Guyot, G. and Major, D. (1989) 'TSAVI: a vegetation index which minimises soil brightness effects on LAI or APAR estimation', in *12th Canadian Symposium on Remote Sensing and IGARSS 1990, Vancouver, Canada, July 10–14*.

Barnett, S. (1990) *Matrices, Methods and Applications*, Oxford: Clarendon Press.

Barrett, E. C. (1970) 'The estimation of monthly rainfall from satellites data', *Monthly Weather Review* 98: 322–7.

Barrett, E. C. (1973) 'Forecasting daily rainfall from satellite data', *Monthly Weather Review* 101: 215–22.

Barrett, E. C. (1974) *Climatology from Satellites*, London: Methuen.

Barrett, E. C. (1979) 'An operational method for rainfall monitoring in north west Africa', Development Project on Remote Sensing Applications for Desert Locust Survey and Control, FAO, Rome.

Barrett, E. C. (1982) ' Development and initial testing of the Bristol InterActive Scheme (BIAS) for satellite improved rainfall monitoring', AgRISTARS Stage 3 Final report to US Department of Commerce contract no. NA-81–SAC-00711.

Barrett, E. C. (1984) 'Development and testing of BIAS-3F: The Bristol/NOAA InterActive Scheme for satellite-improved rainfall monitoring and the BIAS User's Guide', Final report to US Department of Commerce contract no. NA-82–SAC-00083.

Barrett, E. C. (1987) 'Estimation of precipitation from AVHRR and Meteosat data over Africa', in *Proceedings of the Twentieth International Symposium on Remote Sensing of the Environment, Nairobi, Kenya 4–10 December 1986*, Environmental Research Institute, Michigan.

Barrett, E. C. and D'Souza, G. (1986) 'An objective monitoring method for African rainfall based on Meteosat VIS and IR images', in *Proceedings of the Conference on Parameterisation of Land Surface Characteristics, ISLSCP, Rome, 2–6 December 1985, ESA SP-248*, European Space Agency: 313–35.

Barrett, E. C. and D'Souza, G. (1988) 'A comparative study of candidate techniques for U.S. rainfall monitoring operations using Meteorological satellite data', Final Report to the US Department of Commerce under Co-operative Agreement no. NA86AA-H-RA001 (Amendment no. 3).

Barrett, E. C. and Grant, C. K. (1979) 'Relations between frequency distributions of cloud cover over the UK based on conventional observations and imagery from Landsat 2', *Weather* 34: 416.

Barrett, E. C. and Harris, R. (1977) 'Infrared nephanalysis', *Meteorological Magazine* 106: 9–16.

Barrett, E. C. and Harrison, A. R. (1986) 'Rainfall monitoring by Meteosat in Africa', Consultant's Report AGRT. Rome: FAO.

Barrett, E. C. and Martin, D. W. (1981) *The Use of Satellites in Rainfall Monitoring*, London: Academic Press.

Barrett, E. C., Beaumont, M., Richards, T. S. and Power, C. H. (1985a) 'A satellite evaluation of the extreme rainfall event of May 1981 in the Sultanate of Oman', Final report (part 3) to the Sultanate of Oman Public Authority for Water Resources contract no. C/2.

Barrett, E. C., Beaumont, M., Richards, T. S. and Power, C. H. (1985b) 'A satellite evaluation of the extreme rainfall event of August 1983 in the Sultanate of Oman', Final report (part 4) to the Sultanate of Oman Public Authority for Water Resources contract no. C/2.

Barrett, E. C., D'Souza, G., Kidd, C. and Power, C. H. (1987) 'Estimation of precipitation from AVHRR and Meteosat data over Africa', in *Proceedings of the Twentieth International Symposium on Remote Sensing of the Environment, Nairobi, Kenya, December 1986*, Environmental Research Institute, Michigan.

Barrett, E. C., D'Souza, G. and Power, C. H. (1990) 'Satellite rainfall estimation techniques using visible and infrared imagery', *Remote Sensing Reviews* 4, 2: 379–414.

Bellerby, T. J. and Barrett, E. C. (1993) 'Progressive refinement: a strategy for the calibration by collateral data of short-period satellite rainfall estimates', *Journal of Applied Meteorology* 32: 1365–77.

Bellon, A. and Austin, G. L. (1986) 'On the relative accuracy of satellite and rain-gauge rainfall measurements over middle latitudes during daylight hours', *Journal of Climate and Applied Meteorology* 25, 11: 1712–24.

Bellon, A., Lovejoy, S. and Austin, G. L. (1980) 'Combining satellite and radar data for the short-range forecasting of precipitation', *Monthly Weather Review* 108: 1554–66.

Bonifacio, R. (1991) 'Rainfall estimation in Africa using remote sensing techniques', in A. S. Belward and C. R. Valenzuela (eds) *Remote Sensing and Geographical Information Systems for Resource Management in Developing Countries*: 215–33.

Browning, K. A. (1979) 'The FRONTIERS plan: a strategy for using radar and satellite imagery for very short-range precipitation forecasting', *Meterological Magazine* 108: 160–84.

Browning, K. A. and Collier, C. (1982) 'An integrated radar-satellite Nowcasting system in the UK', in K. A. Browning (ed.) *Nowcasting*, London: Academic Press: 47–61.

Bunting, J. T. and Hardy, K. R. (1984) 'Cloud identification and characterisation', in A. Hendersen-Sellers (ed.) *Satellite Sensing of a Cloudy Atmosphere*, London: Taylor and Francis: 203–40.

Bussell, M. A. (1995) 'Software review: TeraVue for Windows, low cost software for remote sensing', *Geological Remote Sensing Group Newsletter* 15: 40–2.

Callis, S. L. and LeComte, D. M. (1987) 'Operational use of satellite imagery to estimate rainfall in the Sahelian countries of Africa', in *Preprints of 18th Conference on Agricultural and Forest Meteorology West Lafayette, American Meteorological Society, Boston, MA, USA*.

Campbell, J. B. (1996) *Introduction to Remote Sensing*, London: Taylor and Francis.

Cappellini, V. (1980) 'Enhancement, filtering and preprocessing techniques', in L. M. Haralick and J. C. Simon (eds) *Issues in Digital Image Processing*, Alphen van den Rijn: Sijthoff and Noordhoff International Publishers B.V.

Carroll, D. M., Evans, R. and Bendelow, V. C. (1977): *Air Photo Interpretation for Soil Mapping*, Soil Survey Technical Monograph 8, Harpenden.

Civco, C. L. (1989) 'Topographic normalization of Landsat Thematic Mapper digital data', *Photogrammetric Engineering and Remote Sensing* 55, 9: 1303–9.

Clark, D. and Borneman, R. (1984), 'Satellite precipitation estimation program for the Synoptic Analysis Branch', in *Proceedings of the Tenth Conference on Weather Forecasting and Analysis, June 25–29 1984, Clearwater Beach, FL*, American Meteorological Society: 392–9.

Clark, D. and Perkins, M. (1985) 'The NOAA Interactive Flash Flood Analyzer', in *Preprints of the International Conference on Interactive Information and Processing Systems for Meteorology, Oceanography and Hydrology, Boston, Mass.* American Meteorological Society: 255–9.

Clevers, J. G. P. W. (1988) 'The derivation of a simplified reflectance model for the estimation of leaf area index', *Remote Sensing of the Environment* 34: 71–3.

Collins, W. (1978) 'Remote sensing and crop type maturity', *Photogrammetric Engineering and Remote Sensing* 44: 4–55.

Condit, C. D. and Chavez, P. S. (1979) 'Basic concepts of computerised digital image processing for geologists', *Geological Survey Bulletin* 1462.

Conover, J. H. (1962) 'Cloud interpretation from satellite altitudes', Research Note 81, Air Force Cambridge Research Laboratories, L. G. Hanscomb Field, MA.

Conover, J. H. (1963) 'Cloud interpretation from satellite altitudes', Supplement to Research Note 81, Air Force Cambridge Research Laboratories, L. G. Hanscomb Field, MA.

Cracknell, A. P. and Hayes, L. W. B. (1991) *Introduction to Remote Sensing*, London: Taylor and Francis.

Creutin, J. D. and Delrieu, J. (1986) 'Remote sensed and ground measurement combination for rainfall field estimation', Paper presented at the Chapman Conference on Modelling of Rainfall Estimation Fields, American Geophysical Union, Caracas, Venezuela.

Creutin, J. D., Lacomba, P. and Obled, C. H. (1986) 'Spatial relationship between cloud-cover and rainfall fields: a statistical approach combining satellite and ground data', in *Proceedings of Hydrologic Applications of Space Technology, Cocoa Beach Workshop, Florida, August 1985*, IAHS Publication no. 160: 81–90.

Crist, E. P. (1983) 'The Thematic Mapper Tasselled Cap – a preliminary formulation', in *Proceedings of Symposium on Machine Processing of Remotely Sensed Data*, West Lafayette: Laboratory for the Applications of Remote Sensing: 357–64.

Crist, E. P. and Cicone, R. C. (1984) 'Application of the Tasselled Cap concept to simulated Thematic Mapper data', *Photogrammetric Engineering and Remote Sensing* 50: 343–52.

Crist, E. P. and Kauth, R. J. (1986) 'The Tasselled Cap demystified', *Photogrammetric Engineering and Remote Sensing* 52: 81–6.

Curran, P. J. (1983) 'Multispectral remote sensing for the estimation of green leaf area index', *Philosophical Transactions of the Royal Society, Series A* 309: 257–70.

Danson, F. M. and Plummer, S. E. (eds) (1995) *Advances in Environmental Remote Sensing*, Chichester: John Wiley and Sons Ltd.

Davis, J. C. (1973) *Statistics and Data Analysis in Geology*, New York: Wiley.

DelBeato, R. and Barrell, S. L. (1985) 'Rain estimation in extratropical cyclones using GMS imagery', *Monthly Weather Review* 113: 747–55.

Dennett, M. D., Elston, J. and Rodgers, J. A. (1985) 'A reappraisal of rainfall trends in the Sahel', *Journal of Climatology* 5: 353–61.

Denniss, A. (1994) 'Dimple – image processing on a Mac', *Geological Remote Sensing Group Newsletter* 11: 40–51.

Desbois, M., Seze, G. and Szejwach, G. (1982) 'Automatic classification of clouds on Meteosat imagery: application to high level clouds', *Journal of Applied Meteorology* 21: 401–12.

Doneaud, A. A., Smith, P. L., Dennis, A. S. and Sengupta, S.

(1981) 'A simple method for estimating the convective rain volume over an area', *Water Resources Research* 17: 1676–82.

Doneaud, A. A., Miller, J. R. Jr, Johnson, L. R., Vonder Haar, T. H. and Laybe, P. (1987) 'The area–time integral technique to estimate convective rain volumes over areas applied to satellite data – a preliminary investigation', *Journal of Climate and Applied Meteorology* 26: 156–69.

Drury, S. A. (1993) *Image Interpretation in Geology*, London: Chapman and Hall.

Drury, S. A. (1998) *Images of the Earth: a guide to remote sensing*, Oxford: Oxford Science Publications.

Drury, S. A. and Berhe, S. M. (1993) 'Accretion tectonics in northern Eritrea revealed by remotely sensed imagery', *Geological Magazine* 130: 177–90.

Drury, S. A. and Walker, A. D. (1987) 'Display and enhancement of gridded aeromagnetic data of the Solway Basin', *International Journal of Remote Sensing* 8, 10: 1433–44.

Dugdale, G. and Milford, J. R. (1986) 'Rainfall estimation over the Sahel using TIR data', in *Proceedings of the Conference on Parameterisation of Land Surface Characteristics, ISLSCP, Rome, 2–6 December 1985*, ESA SP-248: 315–20.

Edwards, K. and Davis, P. A. (1994) 'The use of intensity–hue–saturation transformation for producing colour shaded-relief images', *Photogrammetric Engineering and Remote Sensing* 60, 11: 1369–74.

Elvidge, C. D. and Chen, Z. (1995) 'Comparison of broad-band and narrow-band red and near infrared vegetation indices', *Remote Sensing of the Environment* 54, 1: 38–48.

ESOC (1985) Meteosat-2 Calibration Report (Issue 13) for July–September 1985, MEP/MET, Meteosat Exploitation Project, ESOC Darmstadt, West Germany, October 1985.

Fenner, J. H. (1982) 'Using satellite imagery to estimate winter season rainfall in Europe', in Preprints, 9th Conference on Weather Forecasting and Analysis, American Meteorological Society, Boston, MA: 245–9.

Flasse, S. P. and Verstraete, M. M. (1993) 'Monitoring the environment with vegetation indices: comparison of NDVI and GEMI using AVHRR data over Africa', in *Vegetation, Modelling and Climate Change Effects*, ed. F. Veroustraete and R. Ceulemans, The Hague: SPB Academic Publishing: 107–35.

Flitcroft, I. D., McDougall, V., Milford, J. R. and Dugdale, G. (1987) 'The calibration and interpretation of Meteosat-based estimates of Sahelian rainfall', in *Proceedings of the VIth Meteosat Scientific User's Meeting, Amsterdam 25–27 November 1986*, EUMETSAT, EUM PO1, vol. II.

Follansbee, W. A. and Oliver, V. J. (1975) 'A comparison of infrared imagery and video pictures in the estimation of daily rainfall from satellite data', NOAA Technical Memorandum, NESS 62, Washington, DC: 1–14.

Fox, D., Hindley, D. and Power, C. (1991) 'The determination of areas of peat in satellite imagery', NRSC Ltd, Report DRA SP(91) WP1 06.

Franklin, S. E. and Giles, P. T. (1995) 'Radiometric processing of aerial and satellite remote-sensing imagery', *Computers and Geosciences* 21, 3: 413–23.

Galbiati, L. J. (1990) *Machine Vision and Digital Image Processing Fundamentals*, New York: Prentice-Hall.

Gibson, P. J. (1991) 'An integrated investigation of the Tow Valley fault system, Ireland, with particular reference to remote sensing techniques', Ph.D. thesis, University of Ulster.

Gibson, P. J. (1993) 'Evaluation of digitally processed geophysical data sets for the analysis of geological features in northern Ireland', *International Journal of Remote Sensing* 14: 161–70.

Gibson, P. J. (1999) *Introductory Remote Sensing*, vol. I: *Principles and Concepts*, London: Routledge.

Gibson, P. J. and Lyle, P. (1993) 'Evidence for a major Tertiary dyke swarm in County Fermanagh, Northern Ireland, on digitally processed aeromagnetic imagery', *Journal of the Geological Society*, London 150: 37–8.

Gilabert, M. A., Conese, C. and Maselli, F. (1994) 'An atmospheric correction method for the automatic retrieval of surface reflectances from TM images', *International Journal of Remote Sensing* 15, 10: 2065–86.

Gillespie, A. R. (1980) 'Digital techniques of image enhancement', in B. S. Siegal and A. R Gillespie (eds) *Remote Sensing in Geology*, New York: John Wiley and Sons Inc: 139–226

Gillespie, A. R., Kahle, A. B. and Walker, R. E. (1986) 'Color enhancement of highly correlated images I. Decorrelation and HSI contrast stretches', *Remote Sensing of the Environment* 20: 209–35.

Gillespie, A. R., Kahle, A. B. and Walker, R. E. (1987) 'Color enhancement of highly correlated images II. Channel ratio and "chromaticity" transformation techniques', *Remote Sensing of the Environment* 22: 343–65.

Goldie, N., Moore, J. G. and Austin, E. E. (1958) Upper Air Temperature over the World, *Geophysical Memoirs*, 101, XIII, London: HMSO.

Goward, S. N., Dye, D. G., Turner, S. and Yang, J. (1993) 'Objective assessment of the NOAA global vegetation index data product', *International Journal of Remote Sensing* 4: 3365–94.

Graham, L. A. and Gallion, C. (1996) 'Image processing under Windows NT: a comparative review', *GIS World*, September issue, 36–41.

Green, W. B. (1983) *Digital Image Processing: a systems approach*, New York: Van Nostrand Reinhold Company Inc.

Griffiths, J. F. (ed.) (1972) *Climates of Africa*, World Survey of Climatology vol. 10, ed. H. E. Landsberg, Amsterdam: Elsevier: 193–221.

Hall, E. L. (1979) *Computer Image Processing and Recognition*, New York: Academic Press.

Hall, F. G., Strebel, D. E., Nickeson, J. E. and Goetz, S. J. (1991) 'Radiometric rectification: toward a common radiometric response among multidate, multisensor images', *Remote Sensing of the Environment* 35: 11–27.

Harris, J. R., Neily, L., Pultz, T., Bercha, N. and Slaney, V. R. (1986) 'Principal component analysis of airborne geophysical data for lithologic discrimination using an image processing system', presented at the Twentieth International Symposium on Remote Sensing of Environment, Nairobi, Kenya.

Hendersen-Sellers, A. (ed.) (1984) *Satellite Sensing of a Cloudy Atmosphere: observing the third planet*, London: Taylor and Francis.

Hielkema J. U., Barrett, E. C., Harrison, A. R., Collela, G. and Petricono, A. (1987) 'Operational rainfall monitoring of Africa using low resolution Meteosat observations', in *Proceedings of the VIth Meteosat Scientific User's Meeting, Amsterdam 25–27 November 1986*, EUMETSAT, EUM PO1, vol. II.

Hord, R. M. (1982) *Digital Image Processing of Remotely Sensed Data*, London: Academic Press.

Horgan, G. W. and Critchley, M. F. (1989) 'The use of Thematic Mapper imagery for peatland mapping and mineral exploration in Ireland', in *Proceedings of a Workshop on Earthnet Pilot Project on Landsat Thematic Mapper Applications* held at Frascati, Italy, in December 1987, ESA: 91–7.

Houze, R. A. Jr and Hobbs, P. V. (1982) 'Organisation and structure of precipitating cloud systems', *Advances in Geophysics* 24: 225–315.

Howarth, P. J. and Boasson, E. (1983) 'Landsat digital enhancements for change detection in urban environments', *Remote Sensing of the Environment* 13: 149–60.

Huete, A. R. (1988) 'A Soil-Adjusted Vegetation Index (SAVI)', *Remote Sensing of the Environment* 25: 295–309.

Hussain, Z. (1991) *Digital Image Processing: practical applications of parallel processing techniques*, New York: Ellis Horwood.

Inoue, T. (1987) 'An instantaneous delineation of convective rainfall areas using split window data of NOAA-7 AVHRR', *Journal of the Meteorological Society of Japan* 65: 469–81.

Jensen, J. R. (1996) Introductory *Digital Image Processing: a remote sensing perspective*, Englewood Cliffs, NJ: Prentice-Hall.

Jordan, C. F. (1969) 'Derivation of leaf area index from quality of light on the forest floor', *Ecology* 50: 663–6.

Kaufman, Y. J. (1985) 'The atmospheric effect on the separability of field classes measured from satellites', *Remote Sensing of the Environment* 18: 21–34.

Kaufmann, H. (1988) 'Mineral exploration along the Aqaba–Levant structure by use of TM-data', *International Journal of Remote Sensing* 9: 1639–58.

Kauth, R. J. and Thomas, G. S. (1976) 'The Tasseled Cap, a graphic description of the spectral–temporal development of agricultural crops as seen by Landsat', in *Proceedings of Symposium on Machine Processing of Remotely Sensed Data*, West Lafayette, Purdue University.

Kelly, K. A. (1985) 'Separating clouds from ocean in infrared images', *Remote Sensing of the Environment* 17: 67–83.

Kidd, C. (1997) 'Rainfall monitoring using passive microwave imaging' in *Remote Sensing Society Newsletter* 89: 14–24.

Kilonsky, B. J. and Ramage, C. S. (1976) 'A technique for estimating tropical open-ocean rainfall from satellite observations', *Journal of Applied Meteorology* 15: 972–5.

Knight, D. (1991) 'Growing threats to peat', *New Scientist* 131 (1780): 27–32.

Krajewski, W. F. (1987) 'Co-kriging radar rainfall and rain-gauge data', *Journal of Geophysical Research* 92: 9571–80.

Kriebel, K. T. (1976) 'On the variability of the reflected radiation field due to differing distributions of the irradiation', *Remote Sensing of the Environment* 4: 257–64.

Kriegler, F. J., Malila, W. A., Nalepka, R. F. and Richardson, W. (1969) 'Preprocessing transformations and their effects on multispectral recognition', in *Proceedings of the Sixth International Symposium on remote sensing of the Environment, University of Michigan, Ann Arbor, MI, USA*, West Lafayette, Purdue University: 97–131.

Laughlin, C. R. (1981) 'On the effect of temporal sampling on the observation of mean rainfall', in D. Atlas and O. W. Thiele (eds) *Precipitation Measurements from Space, Workshop Report, October 1981*, NASA Goddard Space Flight Center, D59–66.

Lebel, T. F. (1986) 'A preliminary study of rainfall estimation from PROFS radar and rain-gauge data', Internal Report Environmental Research Laboratory, NOAA, Boulder, CO.

LeComte, D. M., Kogan, F. N., Steinborn, C. A. and Lambert, L. (1988) 'Assessment of crop condition in Africa', NOAA Technical Memorandum, NESDIS AISC 13.

Lee, B. G., Chin, R. T. and Martin, D. W. (1985) 'Automated rain-rate classification of satellite images using statistical pattern recognition', *IEEE Transactions on Geoscience and Remote Sensing* GE-23, 315–24.

Lee, M. K., Pharaoh, T. C. and Soper, N. J. (1990) 'Structural trends in central Britain from images of gravity and aeromagnetic fields', *Journal of the Geological Society, London* 147: 241–58.

Lee, R. and Taggert, C. I. (1969) 'A procedure for satellite-photo-interpretation and appearance of clouds from satellite altitudes', in *Satellite Meteorology*, Bureau of Meteorology, Melbourne.

Leroux, M. (1983) *The Climate of Tropical Africa*, Paris: Champion.

Liljas E. (1982) 'Automated technique for the analysis of satellite cloud imagery', in K. A. Browning (ed.) *Nowcasting*, London: Academic Press: 167–76.

Liljas, E. (1987) 'Multispectral classification of cloud, fog and haze', in R. A. Vaughan (ed.) *Remote Sensing Applications in Meteorology and Climatology*, NATO ASI Series 201: 301–19.

Lillesand, T. M. and Kiefer, R. W. (1994) *Remote Sensing and Image Interpretation*, New York: John Wiley and Sons.

Lira, J. and Oliver, A. (1983) 'A diffusion model to correct multispectral images for the path-radiance atmospheric effect', in R. M. Haralick (ed.) *Pictorial Data Analysis*, Berlin: Springer-Verlag: 385–403.

Liu, C. H., Chen, A. J. and Liu, G. R. (1994) 'Variability of the bare soil albedo due to different solar zenith angles and atmospheric haziness', *International Journal of Remote Sensing* 15, 13: 2531–42.

Lo, C. P. (1986): *Applied Remote Sensing*, London: Longman.

Lovejoy, S. (1981) 'Combining visible and infrared techniques with LAMMR for daily rainfall estimates', in D. Atlas and O. W. Thiele (eds) *Precipitation Measurements from Space, Workshop Report*, October 1981, NASA Goddard Space Flight Center, D59–66, 184–91.

Lovejoy, S. and Austin, G. L. (1979a) 'The delineation of rain areas from visible and infrared satellite data for GATE and mid-latitudes', *Atmosphere–Ocean* 17: 77–92.

Lovejoy, S. and Austin, G. L. (1979b) 'The sources of error in rain amount estimation schemes for GOES visible and infrared satellite data', *Monthly Weather Review* 107: 1048–54.

Lynn, D. W. (1984) 'Surface material mapping in the English Fenlands using airborne multispectral scanner data', *International Journal of Remote Sensing* 5, 4: 699–713.

Maling, D. H. (1973) *Co-ordinate Systems and Map Projections*, London: George Philip and Son Ltd.

Mangolini, M. and Arino, O. (1996) 'ERS-SAR and Landsat-TM multitemporal fusion for crop statistics', *Earth Observation Quarterly* 51: 11–15.

Marion, A. (1991) *An Introduction to Image Processing*, London: Chapman and Hall.

Martin, D. W. and Howland, M. R. (1986) 'Grid history: a geostationary satellite technique for estimating daily rainfall in the Tropics', *Journal of Climate and Applied Meteorology* 25: 184–95.

Mather, P. M. (1987) *Computer Processing of Remotely-Sensed Images*, Chichester: John Wiley and Sons.

Meer, F. van der (1995) 'Spectral unmixing of Landsat Thematic Mapper data', *International Journal of Remote Sensing* 16, 16: 3189–94.

Meisner, B. N. and Arkin, P. A. (1987) 'Spatial and annual variations in the diurnal cycle of large-scale tropical convective cloudiness and precipitation', *Monthly Weather Review* 115: 2009–32.

Middleton, R. (ed.) (1990) 'The North West Wetlands Survey Annual Report', Lancaster University Archaeology Unit.

Milford, J. R. and Dugdale, G. (1987) 'Rainfall mapping over West Africa in 1986', Consultants' Report GCP/INT/432/NET, FAO, Rome, Italy.

Miller, D. M., Kaminsky, E. J. and Rana, S. (1995) 'Neural network classification of remote-sensing data', *Computers and Geosciences*, 21, 3: 377–86.

Milovich, J. A., Frualla, L. A. and Gagliardini, D. A. (1995) 'Environment contribution to the atmospheric correction for Landsat MSS images', *International Journal of Remote Sensing* 16, 14: 2515–37.

Morrison, N. (1994) *Introduction to Fourier Analysis*, New York: John Wiley and Sons Inc.

Moses, J. and Barrett, E. C. (1986) 'Interactive procedures for estimating precipitation from satellite imagery', *Hydrologic Applications of Space Technology: Proceedings of the Cocoa Beach Workshop, Florida August 1985*, IAHS Publication, 160: 25–39.

Motell, C. E. and Weare, B. C. (1987) 'Estimating tropical rainfall using digital satellite data', *Journal of Climate and Applied Meteorology* 26: 1436–46.

Narayana, A., Solanki, H. U., Krishna, B. G. and Narain, A. (1995) 'Geometric correction and radiometric normalisation of NOAA AVHRR data for fisheries applications', *International Journal of Remote Sensing* 16, 4: 765–71.

NCC Scotland (1987) *The Ecology and Management of Upland Habitats: the role of remote sensing*, Special Publication no. 2. Report on one-day workshop held at NCC Scotland HQ, 24 July 1987.

Negri, A. J. and Adler, R. F. (1987a) 'Infrared and visible satellite rain estimation Part I: A grid cell approach', *Journal of Climate and Applied Meteorology* 26: 1553–64.

Negri, A. J. and Adler, R. F. (1987b) 'Infrared and visible satellite rain estimation Part II: A cloud definition approach', *Journal of Climate and Applied Meteorology* 26: 1565–76.

Negri, A. J., Adler, R. F. and Wetzel, P. (1984) 'Satellite rain estimation: an analysis of the Griffith–Woodley technique', *Journal of Climate and Applied Meteorology* 26: 102–16.

Neil, L. K. (1984) 'Estimating precipitation areas over the eastern Pacific using GOES satellite imagery', in Preprints, Conference on Satellite Remote Sensing and Applications, American Meteorological Society, Boston, MA: 121–6.

NOAA (1989) *Storm Data and unusual weather phenomena* vol. 31 no. 9, NOAA National Weather Service, September 1989.

Novak, K. (1992) 'Rectification of digital imagery', *Photogrammetric Engineering and Remote Sensing* 58, 3: 339–44.

Ojo, O. (1977) *The Climate of West Africa*, London: Heinemann.

Pairman, D. and Kittler, J. (1986) 'Clustering algorithms for use with images of clouds', *International Journal of Remote Sensing* 7, 7: 855–66.

Pinty, B. and Verstraete, M. M. (1991) 'Extracting information on surface properties from bidirectional reflectance measurements', *Journal of Geophysical Research* 96: 2865–74.

Power, C. H. (1989) 'Use of satellite remote sensing of cloud and rainfall for selected operational applications in the fields of applied hydrology and food production', Ph.D. thesis, University of Bristol.

Power, C. H., Barrett, E. C. and Beaumont, M. J. (1989) 'A satellite climatology of Arabian Sea tropical storms and cyclones 1966–1987', Final report to the Sultanate of Oman Ministry of Environment and Water Resources contract no. C1/88.

Qi, J., Chenbouni, A., Huete, A. R. and Kerr, Y. H. (1994) 'Modified Soil Adjusted Vegetation Index (MSAVI)', *Remote Sensing of the Environment* 101: 15–20.

Rao, P. K. (1970) 'Estimating cloud amount and height from satellite infrared radiation data', ESSA Technical Report NESC 54, Washington DC.

Ray, T. W. (1996) 'Frequently-Asked Questions (FAQ) on vegetation indices and vegetation in remote sensing', available on the Internet: terrill@mars1.gps.caltech.edu, created by T. W. Ray, Division of Geological and Planetary Sciences, California Institute of Technology, Mail code 170–25, Pasadena, CA 91125, USA.

Reynolds, S. and Vonder Haar, T. H. (1977) 'A bi-spectral method for cloud parameter determination', *Monthly Weather Review* 105: 446–57.

Richards, F. and Arkin, P. A. (1981) 'On the relationship between satellite observed cloud cover and precipitation', *Monthly Weather Review* 109: 1081–93.

Richards, J. A. (1993) *Digital Image Analysis: an introduction*, Berlin: Springer-Verlag.

Richardson, A. J. and Wiegand, C. L. (1977) 'Distinguishing vegetation from soil background', *Photogrammetric Engineering and Remote Sensing* 43: 1541–52.

Robertson, F. R. (1985) 'Infrared precipitation in the South Pacific during FGGE SOP-1', in Preprints, 16th Conference on Hurricanes and Tropical Meteorology, 14–17 May 1985, Houston, American Meteorological Society, Boston, MA.

Sabins, F. F. (1997) *Remote Sensing: principles and interpretation*, New York: W. H. Freeman and Company.

Sader, S. A., Powell, G. V. N. and Rappole, J. H. (1991) 'Migratory bird habitat monitoring through remote sensing', *International Journal of Remote Sensing* 12, 3: 363–72.

Schowengerdt, R. A. (1983) *Techniques for Image Processing and Classification in Remote Sensing*, London: Academic Press Inc.

Scofield, R. A. (1984) 'Satellite-based estimates for heavy precipitation', in *Recent Advances in Civil Space Remote Sensing*, Society of Photo-optical Instrumentation Engineeers, no. 481.

Scofield, R. A. (1986) 'Satellite convective and extra-tropical cyclone cloud categories associated with heavy precipitation', in A. I. Johnson (ed.) *Hydrologic Applications of Space Technology, Proceedings of Workshop, Cocoa Beach, FL*, IAHS Publications 160 47–57.

Scofield R. A. and Oliver, V. J. (1977) 'A scheme for estimating convective rainfall from satellite imagery', NOAA Technical Memorandum, NESS 86, Washington, DC.

Simpson, J. A., Adler, R. F. and North, G. R. (1988) 'A proposed Tropical Rainfall Measuring Mission (TRMM) satellite', *Bulletin of the American Meteorological Society* 69: 278–95.

Skaley, J. E. (1980) 'Photo-optical techniques of image enhancement', in B. S. Siegal and A. R. Gillespie (eds) *Remote Sensing in Geology*, New York: Wiley and Sons: 119–38.

Snijders, F. L. (1991) 'An evaluation of techniques for monitoring of rainfall over the western Sahel using Meteosat PDUS data', in *Proceedings of the 7th Meteosat Scientific User's Meeting, Madrid, 27–30 September 1988*, EUMETSTAT: 275–83.

Snijders, F. L. and Minamiguchi, N. (1998) 'Large area monitoring of crop growing conditions by FAO – ARTEMIS', in *Proceedings of International Symposium on Satellite Remote Sensing for the Earth Sciences, Tokyo, Japan, 5–6 March 1998*.

Stohr, C. J. and West, T. R. (1985) 'Terrain and look angle effects upon multispectral scanner response', *Photogrammetric Engineering and Remote Sensing* 51, 2: 229–35.

Stove, G. C. and Hulme, P. D. (1980) 'Peat resource mapping in Lewis using remote sensing techniques and automated cartography', *International Journal of Remote Sensing* 1, 4: 319–44.

Stove, G. C., Hulme, P. D. and Robertson, R. A. (1980) 'Evaluation of a thermal infrared linescan survey over a hill peatland test area in Scotland', Report for Environment and Resources Consultancy Division, Clyde Surveys Ltd.

Swain, P. H. (1990) 'Image data analysis in remote sensing', in L. M. Haralick and J. C. Simon (eds) *Issues in Digital Image Processing*, Alphen vad den Rijn: Sijthoff and Noordhoff International Publishers B.V.

Teillet, P. M., Guindon, B. and Goodenough, D. G. (1982) 'On the slope-aspect correction of multispectral scanner data', *Canadian Journal of Remote Sensing* 8, 2: 84–106.

Teuber, J. (1993) *Digital Image Processing*, New York: Prentice-Hall International Limited.

Thompson, B. W. (1965) *The climate of Africa*, London: Oxford University Press.

Tsonis, A. A. (1984) 'On the separability of various classes from the GOES visible and infrared data', *Journal of Climate and Applied Meteorology* 23, 10: 1393–410.

Tsonis, A. A. (1987) 'Determining rainfall intensity and type from GOES imagery in midlatitudes', *Remote Sensing of the Environment* 21: 29–36.

Tsonis A. A. and Isaac, G. A. (1985) 'On a new approach for instantaneous rain area delineation in the mid-latitudes using GOES data', *Journal of Climate and Applied Meteorology* 24, 11: 1208–18.

Tucker, C. J. (1979) 'Red and photographic infrared linear combinations for monitoring vegetation', *Remote Sensing of the Environment* 8: 127–50.

Tucker C. J., Townsend, J. R. G. and Goff, T. E. (1985) 'African land cover classification using satellite data', *Science* 227: 369–74.

Turpeinen, O. M., Abidi, A. and Belhouane, W. (1987) 'Determination of rainfall with the ESOC Precipitation Index', *Monthly Weather Review* 115: 2699–706.

UK Meteorological Office (1997) *Nimrod: the leading edge of nowcasting*, Bracknell: Meteorological Office Crown copyright (97/867).

Vincent, R. K. (1997) *Fundamentals of Geological and Environmental Remote Sensing*, Englewood Cliffs, NJ: Prentice-Hall Inc.

Wang, F. (1990) 'Improving remote sensing image analysis through fuzzy information representation', *Photogrammetric Engineering and Remote Sensing* 56, 8: 1163–9.

Weiss, M. and Smith, E. A. (1987) 'Precipitation discrimination from satellite infrared temperatures over CCOPE mesonet region', *Journal of Climate and Applied Meteorology* 26: 687–97.

Whitney, L. F. Jr and Herman, L. D. (1979) 'A statistical approach to rainfall estimation using satellite data', in M. Deutsch (ed.) *Satellite Hydrology*, American Water Resources Association: 139–43.

Wielicki, B. A. and Welch, R. M. (1986) 'Cumulus cloud properties derived using Landsat satellite data', *Journal of Climate and Applied Meteorology* 25, 3: 261–75.

Wilcock, D. N. (1992) 'Image processing and habitat change in the Antrim Coast and Glens Area of Outstanding Natural Beauty', in A. Cooper and P. Wilson (eds) *Managing Landuse Change*, Geographical Society of Ireland Special Publications no. 7.

WMO (1977) 'The use of satellite imagery in tropical cyclone analysis', Geneva: WMO Technical Notes.

World Meteorological Organisation (1986) 'Report of the workshop on global large-scale precipitation data sets for the World Climate Research Programme', Camp Springs, USA, 24–26 July 1985, WCRP-1182.

Wu, R., Weinman, J. and Chin, R. (1985) 'Determination of rainfall rates from GOES satellite images by a pattern recognition technique', *Journal of Atmospheric and Oceanic Technology* 2: 314–30.

Wylie, D. P. and Laitsch, D. (1983) 'The impacts of different satellite data on rain estimation schemes', *Journal of Climate and Applied Meteorology* 22: 1270–81.

PRACTICAL MANUAL FOR *DIGITAL IMAGE PROCESSING*

See also the linked Website to accompany this CD-ROM at:
www.remote-sensing.routledge.com

PRACTICAL MANUAL FOR DIGITAL IMAGE PROCESSING

FOR USE WITH THE ACCOMPANYING *INTRODUCTORY REMOTE SENSING* CD-ROM

Manual Outline

1 Introduction to DRAGON Software
2 CD-ROM content
3 Installation
4 Functions and operation
5 Datasets for practical exercises
6 Practical exercises

---

1 INTRODUCTION TO 'DRAGON' IMAGE PROCESSING SOFTWARE

The software provided on the accompanying CD is a Limited Edition of DRAGON which allows the processing of remotely sensed images using IBM and IBM-compatible personal computers. A full functionality version of DRAGON software can be purchased from Goldin-Rudahl Systems Inc. (see details at end of book).

The software will allow you:

- to display, manipulate, measure and analyse remotely sensed images in single-band format

and as three-band combinations in false colour composites;

- to perform different types of contrast stretching on the imagery, including user-defined stretches;
- to zoom in to look at the images in more detail and to find out the digital values associated with specific positions on the images;
- to produce DN histograms that summarise the distribution of the digital values in one band;
- to produce scatterplots of two bands which provide a qualitative view of the correlation between bands;
- To measure linear and areal features on the images with different units of measure;

- To produce ratio images;
- To apply filters to the imagery, including user-defined ones;
- To perform unsupervised and supervised classification;
- To save the results of any processing to a separate file.

Despite the variety of operations that DRAGON can perform, it is an easy-to-learn package, being mainly mouse-driven with a user interface based on menus with items to select and response panels with fields to complete. DRAGON Limited Edition handles digital images as separate files. Therefore, in the case of multispectral images, each band is a separate image file. Thus, to create a false colour composite you will be required to select the different band files that you wish to combine for display.

2 CD-ROM CONTENT

The files on the CD-ROM are organised as follows:

\IMAGES contains all of the sample images and related data used in the textbook exercises.
SETUP.EXE is a Windows program you can use to install DRAGON onto your hard disk. You can NOT run DRAGON directly from the CD-ROM.
README.TXT provides instructions for installation, information about the computer requirements for installing the Limited Edition, information about the differences between the Limited Edition and the full retail edition, and some hints in case you encounter difficulty in installing the Limited Edition.
\WEBSITE provides complete technical specifications, plus some demonstration images, and links to external Web sites taken from the Goldin–Rudahl Systems, Inc. World Wide Web website.

There is no DRAGON software users' manual provided with the DRAGON Limited Edition, either printed or on the CD-ROM . However, DRAGON has extensive Help available within the program,

which you can access by pressing the F4 key, or by clicking on the [?] symbol.

3 INSTALLATION

To install the DRAGON Limited Edition from within Windows, simply run the setup program. You may either:

- in Windows 3.1, go to the File Manager, click on File/Run/Browse and select first your CD-ROM drive, then the setup.exe progam at the top-level.
- in Windows 95 or (WindowsNT), click on Start/Run/Browse, and select first your CD-ROM drive, then the setup.exe progam at the top level.
- Alternativeley in Windows 95 or (WindowsNT), open a DOS box, make your CD-ROM drive as your current drive then type setup.

Even though you can install DRAGON from inside Windows, you must remember that the Limited Edition is really an MS-DOS program, and is running either in MS-DOS alone, or as a full screen MS-DOS box within Windows. You should not expect to be able to use the Windows clipboard, for example, in conjunction with the Limited Edition.

Of course, you do not need to use Windows to install the DRAGON Limited Edition. If you are in DOS, simply change to your CD-ROM drive and type dossetup.

IMPORTANT NOTE: Even after it has been installed, DRAGON Limited Edition will not run unless the CD-ROM remains in your CD-ROM drive.

4 FUNCTIONS AND OPERATION

DRAGON Limited Edition consists of a main menu which gives access to its six main functions:

- Display images
- Enhance images

- Classify images
- Geometry operations
- Utility operations
- File handling operations

Each of these functions has its own dedicated menu. (One may also find out more about or exit DRAGON from the Main Menu.) From the Main Menu, once the desired option is selected, the user is brought to the menu dedicated to that function and has the choice of accepting the default options or of altering the operations on the images.

DRAGON also has an on-line Help facility which is always accessible from any menu by clicking with the mouse on the ? symbol. Therefore, it is always possible and easy to interrogate the package about details and instruction on how to carry out the desired operation on the image. In addition, DRAGON Limited Edition makes the main controls – Return to previous menu, Top level of the menu system, and eXit – available even if they are not specifically listed in the menu, by typing R, T and X respectively. In addition, the ESC key will bring you back to the previous menu. Moreover, all the single-letter commands (Display, Enhance, etc.) are available at every menu even if not listed, by typing in the appropriate letter.

Initiation of the program software produces the Main Menu. This can be accomplished by clicking on the icon or if in MS-DOS mode by typing DRAGON while in the DRAGON subdirectory. The Main Menu gives access to all the principal functions of DRAGON Limited Edition. These can be selected by clicking with the left-hand mouse button on the corresponding line on the screen or typing the letter at the beginning of the line (D, for example, opens the Display menu) and then pressing the RETURN key on the keyboard.

Display

The Display function consists of different options:

-1BA-	Display a single image file in colour
-3BA-	Combine three different image band files,

displaying them in red, green and blue respectively, in order to create a false colour composite image

-GRA-	Display a single-image band file in shades of grey.
-R-	Return to previous menu
-X-	Exit DRAGON Limited Edition

Once the desired option has been selected, the screen will display a different menu for each of the above options which require an input from the user. Details of these inputs are given in the practicals contained in this section.

Enhancement

The Enhancement function consists of six different options:

-SUM-	Sum of two image bands
-DIF-	Difference of two image bands
-RAT-	Ratio of two image bands
-FIL-	Filtering applied to an image
-R-	Return to previous menu
-X-	Exit DRAGON Limited Edition

The use of these options is detailed within the practicals.

Classify

The Classify option consists of two main classification methods, supervised and unsupervised, Minimum Distance to Mean and Clustering respectively. The selection of the Classify option gives access to its dedicated menu and its control options. The function menu consists of the following operations:

-TRA-	Training signature creation
-LIS-	List signature in memory or file
-EDI-	Edit training signature
-MDM-	Minimum Distance to Mean
-CLU-	Clustering
-R-	Return to previous menu
-X-	Exit DRAGON Limited Edition

Geometry

The Geometry function menu includes operations that allow you to measure line distances and areas using the image as a background and the mouse cursor as a tracing tool. DRAGON Limited Edition displays the measurements as number of pixels by default. However, it is possible to select the most appropriate unit of measure. The menu of this function includes the following options:

-MEA-   Measure lengths and areas
-R-   Return to previous menu
-X-   Exit from DRAGON Limited Edition

Utility

This function includes operations that perform a variety of useful tasks on the images and the respective files. The Utility function allows the user to list the statistical, historical and identification information stored in the image files headers. It is also possible to display the cursor co-ordinates and data values at any point on the screen, display the histogram of a single band, and produce scatterplots for two bands. The menu of the Utility functions consists of the following operations:

-LIS-   List header information for image files
-CUR-   Display cursor co-ordinates
-HIS-   Display the histogram of the image
-SCA-   Display the scatterplot of two image bands
-R-   Return to previous menu
-X-   Exit from DRAGON Limited Edition

File Handling Operations

The File handling menu allows the operator to save the currently displayed image as a new image, thus saving the effects of a particular processing such as a filter operation. The menu of the File handling operations function consists of the following options:

-SAV-   Save current memory image in file
-R-   Return to previous menu
-X-   Exit from DRAGON Limited Edition

5 DATASETS FOR PRACTICAL EXERCISES

Datasets for nine scenes are provided with the DRAGON image processing software in order that some of the processing techniques and applications discussed in the text can be performed. It is not possible to produce Intensity/Hue/Saturation images or principal component images with the version of DRAGON that is provided. However, these data have been produced separately and converted to DRAGON format and are provided for some of the scenes. In all, 77 datasets are provided from different parts of the globe from different sensors: SPOT, TM and radar. Using these data in various combinations, it is possible to create literally thousands of different images. The datasets within the D:\images\ subdirectory are:

Northwest Ireland Dataset

This dataset comprises 15 bands of data for the Donegal region of northwest Ireland: six reflected TM bands; six principal components and Intensity, Hue and Saturation images. The scene is 400 pixels by 400 lines in size (12 × 12 km) and is centred on 54 degrees, 50 minutes north and 8 degrees, 12 minutes west. This is an upland region characterised by heather- and gorse-covered granite in the northwest and older metamorphic rocks to the east. There is little evidence for cultivation though managed forestry is carried out. The largest town in the region is Glenties, Figure M1.

Don1.img	TM band 1
Don2.img	TM band 2
Don3.img	TM band 3
Don4.img	TM band 4
Don5.img	TM band 5
Don7.img	TM band 7
Donpc1.img	First principal component
Donpc2.img	Second principal component
Donpc3.img	Third principal component
Donpc4.img	Fourth principal component
Donpc5.img	Fifth principal component
Donpc6.img	Sixth principal component

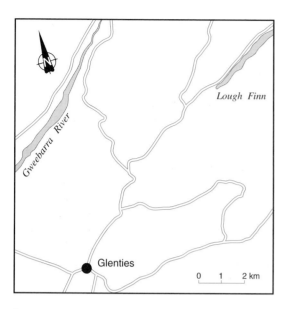

Figure M.1 Location map for Northwest Ireland dataset.

Doni.img	Intensity image
Donh.img	Hue image
Dons.img	Saturation image

Nevada Dataset

This dataset comprises 13 bands of data for the Salt Wells part of Nevada: five reflected TM bands, five principal components and Intensity, Hue and Saturation images. The scene is 400 pixels by 400 lines in size (approximately 12 × 12 km) and is centred on 39 degrees, 30 minutes north and 118 degrees, 31 minutes west. The scene is dominated by three main surface classes: agricultural patterns in the northwest, a lake in the northeast and salt-pans and superficial sediments in the south.

Nev1.img	TM band 1
Nev3.img	TM band 3
Nev4.img	TM band 4
Nev5.img	TM band 5
Nev7.img	TM band 7
Nevpc1.img	First principal component
Nevpc2.img	Second principal component
Nevpc3.img	Third principal component
Nevpc4.img	Fourth principal component
Nevpc5.img	Fifth principal component
Nevi.img	Intensity image
Nevh.img	Hue image
Nevs.img	Saturation image

China Dataset

This dataset comprises six bands of data of the Kowloon region of China. The scene is 400 pixels by 400 lines in size (8 × 8 km) and is centred approximately on 22 degrees, 19 minutes north and 114 degrees, 12 minutes east. The first three bands are SPOT data obtained when the sensors were pointing directly beneath the satellite (nadir viewing). The second set of three SPOT bands cover the same region but were obtained off-nadir when the sensors were pointing at an angle of 27.5 degrees. A conurbation dominates the southern part of the scene whereas the northern part represents mainly natural vegetation.

Ch1(n).img	Multispectral SPOT band 1 (nadir view)
Ch2(n).img	Multispectral SPOT band 2 (nadir view)
Ch3(n).img	Multispectral SPOT band 3 (nadir view)
Ch1(o).img	Multispectral SPOT band 1 (oblique view)
Ch2(o).img	Multispectral SPOT band 2 (oblique view)
Ch3(o).img	Multispectral SPOT band 3 (oblique view)

Virginia, Ireland, Dataset

This dataset comprises 15 bands of data for the Virginia region of Ireland: six reflected TM bands, six principal components and Intensity, Hue and Saturation images. The scene is 400 pixels by 400 lines in size (12 × 12 km) and is centred on 53 degrees, 51 minutes north and 7 degrees, 8 minutes west. This is a lowland agricultural region characterised by small, irregularly shaped fields. Lough Ramor dominates the southern part of the image. The largest town in this region, Virginia, lies on the northern shore of Lough Ramor and is connected to a settlement, Ballyjamesduff, by the N3 road (Figure M2).

V1.img	TM band 1
V2.img	TM band 2
V3.img	TM band 3
V4.img	TM band 4
V5.img	TM band 5
V7.img	TM band 7
Vpc1.img	First principal component
Vpc2.img	Second principal component
Vpc3.img	Third principal component
Vpc4.img	Fourth principal component
Vpc5.img	Fifth principal component
Vpc6.img	Sixth principal component
Vi.img	Intensity image
Vh.img	Hue image
Vs.img	Saturation image

Sudan Dataset

This dataset comprises three bands of radar data for an area of north Sudan. The data were obtained by the Space Shuttle during the SIR-C experiment in April 1994. The scene is 400 pixels by 400 lines in size (4.5 × 4.5 km) and is centred on 19 degrees, 30 minutes north and 33 degrees, 30 minutes east. This region lies immediately to the east of the River Nile. Bare rock is exposed in the west whereas the eastern half of the image is sand-covered. However, radar can penetrate the sand to some extent and reveal geological structures beneath it. The NNW–SSE-trending lineament is a fault. Irregular lines represent small dried-up river channels (wadis).

Sudan1.img	Radar image (L band: vertical transmit vertical return)
Sudan2.img	Radar image (L band: horizontal transmit horizontal return)
Sudan3.img	Radar image (C band: horizontal transmit vertical return)

Andes Dataset

This dataset comprises 15 bands of data for part of the Andes mountain range in Peru: six reflected TM bands, six principal components and Intensity, Hue and Saturation images. The scene is 400 pixels by 400 lines in size (12 × 12 km) and is centred on 13 degrees, 20 minutes south and 75 degrees, 50 minutes west. The topography varies by 2,000 m for this region and it is characterised by two rivers, the main one being the Rio San Juan.

Peru1.img	TM band 1
Peru2.img	TM band 2
Peru3.img	TM band 3
Peru4.img	TM band 4
Peru5.img	TM band 5
Peru7.img	TM band 7
Perupc1.img	First principal component
Perupc2.img	Second principal component
Perupc3.img	Third principal component
Perupc4.img	Fourth principal component
Perupc5.img	Fifth principal component
Perupc6.img	Sixth principal component
Perui.img	Intensity image
Peruh.img	Hue image
Perus.img	Saturation image

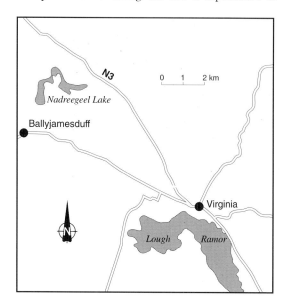

Figure M.2 Location map for Virginia dataset.

Rhine Dataset

This dataset comprises three bands of data for the Rhine floodplain on the border between Germany and France. The scene is 400 pixels by 400 lines in size (8 × 8 km) and is centred approximately on 47 degrees, 59 minutes north and 7 degrees, 38 minutes east. This region is intensively cultivated and is characterised by regular field patterns, many of which contain crops or pasture, though some have bare soil at the surface.

Rhine1.img	Multispectral SPOT band 1
Rhine2.img	Multispectral SPOT band 2
Rhine3.img	Multispectral SPOT band 3

East Ireland Dataset

This dataset comprises 15 bands of data for the area around Maynooth, Ireland: three reflected TM bands, three radar bands, six principal components and Intensity, Hue and Saturation images. The scene is 400 pixels by 400 lines in size (12 × 12 km) and is centred on 53 degrees, 22 minutes north and 6 degrees, 35 minutes west. This is a lowland agricultural region characterised by small fields. Three towns (Maynooth, Celbridge and Leixlip) and a large estate (Carton) are within the study region (Figure M3).

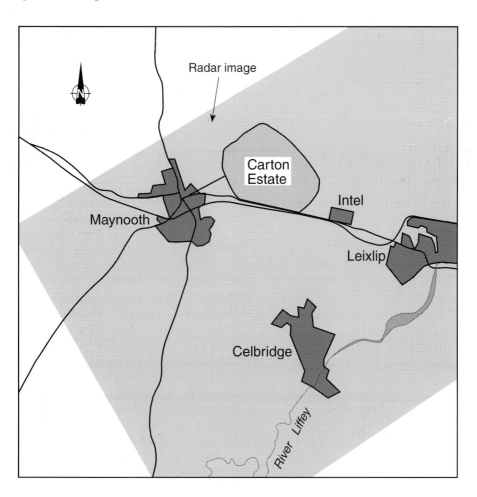

Figure M.3 Location map for East Ireland dataset.

May3.img   TM band 3
May4.img   TM band 4
May5.img   TM band 5
Mayr1.img   L-band radar image (wavelength 23.5 cm) HH polarisation
Mayr2.img   L-band radar image (wavelength 23.5 cm) VV polarisation
Mayr3.img   C-band radar image (wavelength 5.8 cm) VV polarisation
Maypc1.img   First principal component
Maypc2.img   Second principal component
Maypc3.img   Third principal component
Maypc4.img   Fourth principal component
Maypc5.img   Fifth principal component
Maypc6.img   Sixth principal component
Mayi.img   Intensity image
Mayh.img   Hue image
Mays.img   Saturation image

United Kingdom Dataset

This dataset comprises seven Landsat TM bands of data. The scene is 362 pixels by 259 lines in size (approximately 11 × 8 km) and is centred on 50 degrees, 35 minutes north and 4 degrees, 5 minutes west. The town of Tavistock is in the southwest part of the scene. Pastoral agricultural land in the west grades northwards and eastwards to open moorland (Dartmoor). Original image data acquired by the European Space Agency in 1988 and distributed by Eurimage/NRSC.

UK1.img   TM band 1
UK2.img   TM band 2
UK3.img   TM band 3
UK4.img   TM band 4
UK5.img   TM band 5
UK6.img   TM band 6
UK7.img   TM band 7

6 PRACTICAL EXERCISES

Practical Exercise 1
Investigation of the spectral characteristics of various surface types on single-band imagery

This practical introduces the commonest enhancement performed on remotely sensed data, namely contrast stretching. The reader should be familiar with the concepts and terminology of section 3.2 of this volume and also the concept of the DN histogram discussed in section 2.2. The aims of this practical are:

- to provide hand-on experience in contrast stretching;
- to understand how choosing different stretches highlights different features;
- to examine the effects of a stretch on the DN histogram;
- to investigate the spectral characteristic of various surface classes at different wavelengths.

After displaying each image you are asked to describe it. A short description is provided in italics but you are encouraged to work through the exercises carefully, making appropriate notes as you progress. The duration of this practical is approximately six hours. The Virginia dataset is provided as a worked example in order to familiarise you with the DRAGON software. You should repeat this exercise using other datasets. **As you will be required to display and often contrast stretch images in later practicals, it is important that you fully work through this practical because you will need to apply the techniques discussed here later on.**

Virginia dataset files to be used:

    D:\images\V1.img
    D:\images\V2.img
    D:\images\V3.img
    D:\images\V4.img
    D:\images\V5.img
    D:\images\V7.img

Switch on the computer and start the DRAGON software by double clicking the left button of the mouse on the icon. (You may also start DRAGON in DOS mode. From the C:\> prompt, type CD DRAGON and then type DRAGON at the C:\DRAGON> prompt.) This will bring you into the DRAGON Main Menu; note that now the cursor has the shape of a small dragon. (If your mouse does not respond, it is probably not set up correctly; try adding the line c: \mouse\mouse to your autoexec.bat and reboot your computer.) Alternatively use the up and down arrow keys ↑↓.

From the DRAGON Limited Edition Main Menu, choose D (Display images) by moving the cursor on it and clicking with the left button of the mouse. Alternatively, use the up and down arrow keys ↑↓ or type the letter D in the command line. Further options will now be displayed on the monitor. Choose GRA (Grey-scale image display), again by moving the cursor and clicking with the left mouse button. These keystrokes have revealed a screen of command lines in green with inputs displayed in yellow. Follow the instructions in BOX A.

Following the above instructions, the image file D:\images\V1.img is displayed on the screen. Note how dark the image is. Describe what you can see (probably not too much! Do not adjust the brightness of your monitor).

The image is dark grey. One might just discern a pale grey lineament running from northwest to southeast (it is the N3 road) but apart from this, little information can be obtained.

Image file D:\images\V1.img (Landsat TM band 1; bandwidth 0.45–0.52 μm)

The image file D:\images\V1.img is acquired in band 1, which encompasses part of the blue portion of the visible range of the electromagnetic spectrum (see Figure 3.2, *Introductory Remote Sensing: Principles and concepts*). Blue light has the shortest wavelength within the visible portion of the spectrum and this makes it vulnerable to scattering by particles in the atmosphere. In particular, the shorter wavelengths (therefore the blue light) are affected by Rayleigh scattering. Even though this band is sensitive to scattering, the

BOX A

You now need to input the name of a file in the first line on the screen, the Image File line. This can be accomplished in either of two ways. The file name, in this case V1.img, can be typed where the cursor is flashing because the default directory, D: \images\, is already given where D: represents the CD-ROM drive. (If you decide to move the files from this subdirectory, you must always input the entire path on the Image File line. A subdirectory C: \DRAGON\images\ is provided.) When you have entered the file name, press the RETURN key. (This may be termed ENTER on your computer.)

An alternative means of inputting the file name is by moving the cursor to cover the blue triangle at the end of the Image File line and clicking the left button. A blue window will appear on the screen that lists the available files. Choose the file V1.img by moving the cursor on it and clicking with the left button. If you cannot see the name of the file, click on the scroll arrows at the top and bottom of the right-hand side of the window to search for the image file. When you have selected the file, click on it and then press the RETURN key. Note that the brown line has now moved below the Histogram Adjustment line, the second line on the screen. Cover the blue triangle at the

right end of the line with the cursor and click the left mouse button. A list of adjustments will appear. Choose NONE by clicking on it with the left-hand mouse button and then move the cursor to the bottom left of the screen to cover RUN and again press the left-hand mouse button.

Function Keys

A number of function keys (the buttons on the top row of the computer) may also be used to accomplish the above.

F1: This is equivalent to the RUN command.

F2: Allows the user to switch between Menu and Command mode.

F3: This button provides a list of possible inputs for that field. Thus, if it is pressed on the Image File line, it shows the available images.

F4: This provides a brief explanation of the current field.

F5: This is similar to F1 except that the new image or display is superimposed on what was previously displayed.

F6: This gives information on the current system status.

F10: This is similar to the CANCEL button which cancels the operation and returns you to the Main Menu.

---

blue radiation is used in remote sensing for its water-penetrative capability and may prove useful in coastal studies.

In order to view the histogram of the image, follow the instructions in BOX B.

Describe the histogram.

Even though the pixels encompass the 0–143 range, most are concentrated about a DN value of 66. The histogram shows no spacing in the most populated pixel classes. The narrow DN range, in which most pixels lie, accounts for the low contrast of the image.

To obtain statistical information about the image file, you can use the LIS (List header) option in the Utility menu. Press ESC to remove the histogram from the screen then left click LIS from the Utility menu. D:\images\V1.img is still held in the memory ($=$M), click RUN and information on the image file will be displayed.

For D:\images\V1.img, the minimum value is 0; the maximum is 143; 66 is the mean value and the standard deviation is 5.919. Click OK to return to the Utility menu.

In order to obtain more information on the image, it is necessary to stretch it. This will initially be

---

BOX B

- Press any key to RETURN to the Display menu and then from the menu, choose R (Return to previous menu) by clicking it with the left button of the mouse (or use the up and down arrow keys ↑↓).
- Now choose U (Utility options) and then HIS (Histogram of the image) when the new display appears by left clicking the mouse.
- Note that the Image File line states = M, this means that the image file D:\images\V1.img is still stored in the computer memory and does not have to be re-entered. Press RETURN.
- The second line on the screen asks if you want to apply a stretch. The default option is NO (no stretch is applied to the histogram). Leave unchanged.
- To display the histogram, move the cursor to the bottom left of the screen and click RUN or F1.

---

accomplished by displaying the image using the LINEAR AUTOMATIC contrast stretch option. Move the cursor onto R and click the mouse button to go back to the Display menu. Move the cursor onto D (Display images) and click the mouse button. Choose GRA (Grey-scale image display) on the new screen and click again.

A screen similar to that displayed before appears but the Image File line now states =M. This means that the image file D:\images\V1.img is still stored in the computer memory. Press RETURN. The LINEAR AUTOMATIC option from the Histogram Adjustment line is given as default. Accept this. Move the cursor to the bottom left of the screen to cover RUN and again press the left-hand mouse button. (If you had wished to choose another histogram adjustment, you would have moved the cursor onto the blue triangle and clicked the left mouse button. Another blue window would have appeared

on the screen. You select the desired option by moving the cursor on it and then clicking with the left button.)

Describe the image after the linear automatic contrast stretch has been applied.

The image is now brighter and the road network is more visible. You might also discern a dark patch to the west of the N3 road in the northwest quadrant of the image, which is a lake.

We can now view and describe the histogram after the LINEAR AUTOMATIC stretch is applied. Press any key to RETURN to the Display menu and then choose R (Return to previous menu) by clicking it with the mouse.

Choose U (Utility options) and then HIS (Histogram of the image) when the new display appears.

The Image File line displays =M because D:\images\V1.img is still in the computer memory.

Note that the second line on the screen asks if you want to apply a stretch. Press the RETURN key or click on the line to make it active. Move the cursor to cover the blue triangle at the end of the line, click with the left button and this time select YES from the blue window, moving the cursor to cover it; click it. To display the histogram move the cursor to the bottom left of the screen and click RUN.

Describe the histogram.

The stretched histogram is now shifted to the right: the peak of the distribution of the pixels across the DN range is at DN 204. The pixel classes are now distributed in the interval 0–255 of the DN output range.

Note that it could be useful to compare the histogram before and after any stretch has been applied, in order to have a better understanding of the effects of the different stretch options. To display again the unstretched histogram, after you have displayed the stretched one, press the ESC key to go back to the Utility menu, select HIS again and press RETURN.

=M means the file remains in the memory; accept this by pressing RETURN. Choose the default NO option at the end of the Apply Stretch line and then press the function key F5. You can now examine the stretched and unstretched histograms together. Pressing the ESC key will bring you back to the Utility menu.

We want now to display the image using the USER-DEFINED stretch.

You will be required to set upper and lower break limits to stretch the image and will be required to enter a value between 0 and 255 for each of the breaks. Setting these values will assign a DN of 255 (white on the image) to all the pixels having a DN value above the upper value you set, and will set to 0 (black on the image) all the pixels having a DN value below the lower break value. Try with different upper and lower break values. Note the effect of different breaks and describe the best resulting image. Press ESC to return to the Main Menu and click on D the Display menu and click on GRA. The Image File line still displays =M.

Press RETURN or click on the Histogram Adjustment line with the mouse to make it active. Then click on the blue triangle at the end of it and select USER-DEFINED from the window that appears on the screen by clicking on it and press RETURN. The screen will display a window with lines for the upper and lower break limits. Enter the upper break value after deleting the default value of 255 by using the backspace key ← and typing the value you chose, then press RETURN.

Enter the lower break value after deleting the default value of 0 by using the backspace key ← and typing the value you chose, then click OK on the lower left corner of the window. Then click on RUN at the bottom left of the screen to display the image with the stretch you defined.

Note that if you are not satisfied with the result obtained from the break limits you set you can try different limits. Press ESC, then GRA at the new screen. The Image File line will display =M, press RETURN to make the Histogram line active and click on the blue triangle; again select USER-DEFINED, click it and then enter new upper and lower break limits.

Good results are obtained by setting the following upper and lower break limits:
Upper break 70
Lower break 60

The image has quite a good contrast. This short-wavelength image is affected by haze, shown as white patches, but the roads and lakes appear much more clearly than on previous images.

You should now repeat the above procedures for the following files within this dataset, noting the appearance of different features at different wavelengths. Note the line =M on the image file line has to be deleted by using the backspace key ← and the next file input including path: D:\images\vir2.img.

D:\images\V2.img image file (Landsat TM band 2; bandwidth 0.52–0.60 mm)

Landsat TM band 2 is within the green range of the visible part of the electromagnetic spectrum. The wavelength is slightly longer with respect to band 1 and this should make it somewhat less sensitive to Rayleigh scattering than TM band 1. Landsat TM band 2 is used for the measurement of the peak of green reflectance from vegetation, for vegetation discrimination and for the assessment of vegetation vigour. Note similarities and differences between TM band 1 and TM band 2 image files of the Virginia dataset.

Unstretched image of D:\images\V2.img
The image is very dark and virtually nothing can be observed.

Unstretched histogram of D:\images\V2.img
Most of the pixels are concentrated in the classes around a DN value of 10 (min. DN 0, max. DN 75), which explains the overall dark grey tone of the image.

Linear automatic stretch of D:\images\V2.img
The image is now much brighter. The largest lake in the region (Lough Ramor, Figure B2), which could not be

detected on D:\images\V1.img, is now visible at the bottom right of the image and a number of smaller lakes can also be seen. The signature for Lake Ramor is different from the lakes in the northwest, possibly because of a thin cloud cover or the lakes might be of different depths. The road network is well displayed with the exception of the portion of the N3 in the upper left of the image. There are a number of bright small dots (small settlements) scattered across the area though the town of Virginia is poorly expressed. The settlement of Ballyjamesduf (middle of the left margin of image) is visible as a bright small area. There are dark patches near the right margin, which could be forested areas since they generally show a lower reflectance than cultivated areas. The field pattern is still poorly expressed.

Histogram of linear automatic stretch applied to D:\images\V2.img

The pixel classes now range from 0 to 255; the most populated class is around a DN value of 144. The wider range of the distribution of the pixel classes determines the higher degree of contrast of the image and therefore its improved legibility.

User-defined stretch applied to D:\images\V2.img
Given the shape of the unstretched histogram, good results are obtained by setting the upper break at 17 and the lower break at 5. The town of Virginia on the northern shore of Lough Ramor, and Ballyjamesduff, are shown as bright tones. The small lakes are visible as well.

User-defined histogram of D:\images\V2.img
After the stretch, the histogram has a triangular shape; all the classes are well separated. In addition, these limits have allowed us to maintain a very good overall discrimination, retaining almost all the information (if the stretch should result in too many pixels set to 0 or 255 there would be a loss of information about the different reflectance values from ground features). The most populated class has a DN of 85.

Image file: D:\images\V3.img (Landsat TM band 3; band width 0.60–0.69 μm)

Landsat TM band 3 is in the visible red portion of the electromagnetic spectrum and should be less suscep-

tible to Rayleigh scatter and produce a sharper image. Note whether you can detect any differences from the D:\images\V1.img and D:\images\V2.img image files.

Unstretched image of D:\images\V3.img
Again, the unstretched image is dark, with poor contrast, and does not appear better than earlier images.

Unstretched histogram of D:\images\V3.img
As expected from the tone of the image, most of the pixels have low DN values: most DNs are lower than 30. The distribution ranges from DN 0 to DN 87; the mean value is 10.

Linear automatic stretch of D:\images\V3.img
The image is dominated by a medium grey tone. The road network and the settlements are visible. Lough Ramor is not too clear, though the small lakes at the top of the image are well defined. Field patterns may be discerned northeast of Lough Ramor.

Histogram of linear automatic stretch applied to D:\images\V3.img
The histogram shows the most populated class at about 88. The pixel classes are distributed along the whole DN range; the histogram is clearly asymmetric, most of the pixels having a DN of less than 128.

User-defined stretch applied to D:\images\V3.img
Reasonably good results are obtained setting the upper break at 13 and the lower break at 6. The image represents the best trade-off between contrast and legibility; all the ground features – road network, settlements, field boundaries and smaller lakes – are visible.

User-defined histogram of D:\images\V3.img
The histogram shows an even distribution of all the pixel classes across the whole DN range. There are fewer pixels with higher DNs.

D:\images\V4.img image file (Landsat TM band 4; bandwidth 0.75–0.90 μm)

Landsat TM band 4 corresponds to near-infrared radiation and cannot be detected by the human visual system. The much longer wavelength compared to TM bands 1 and 2 makes it far less sensitive to scattering, and therefore it has good atmospheric penetration capabilities. As vegetation has a very high reflectance in the near infrared compared to the visible range, TM band 4 is particularly well suited to vegetational studies. In addition, because of the strong absorption of near infrared from water, TM band 4 can be used for water body delineation and for soil moisture discrimination (areas with the higher moisture content will tend to have a lower reflectance and therefore a darker signature).

Unstretched image of D:\images\V4.img
This image is completely different from the images displayed so far. All the water features are visible (black) and water–land boundaries are clearly defined. New water features are now visible such as a small lake connected by a stream to Lough Ramor. There are several other small lakes visible in the upper right quadrant of the displayed area.

The overall tone of the image is bright and the features are shown with greater definition: the field pattern is more evident and there is less chance of misinterpreting dark vegetation patches as lakes.

The road network is quite well defined but settlements are not distinctive and generally have a mid-grey signature. However, their location can be inferred at the junctions of the roads

Unstretched histogram of D:\images\V4.img
The histogram reflects the differences between this image and the previous ones: the major part of the information is concentrated between DN values of 64 and 128, but the pixel classes range from 0 up to 163 with a mean value of 93. (If one uses the LIS facility in the Utility menu, one can determine that the standard deviation for this band (29.20) is much greater than for the previous bands. This reflects the large range of DNs for vegetation in the infrared). In addition, the histogram displays two different peaks: a smaller one with DN values between 0 and 20 (centred on 10 approximately) is most likely due to the water features of the image; the major one has its peak at DN 108 and ranges from 32 to 140 (approximately).

Linear automatic stretch of D:\images\V4.img
The contrast for the image has increased somewhat and the field patterns displayed in shades of grey are visible. The large irregular mid-grey tone on the northern shore of Lough Ramor, west of Virginia, is probably forest.

Histogram of linear automatic stretch applied to D:\images\V4.img
The shape of this histogram maintains the two separate peaks. However, it is no longer blocky: the spacing among the different classes of the distribution is now wider and they now cover the whole DN output range 0–255, the peak being at DN 170. The main difference from the histograms of the previous bands is that there are far more classes of pixels, providing greater detail. Note that the number of pixels per class is now greatly reduced. The vertical axis of the plot has a maximum value (number of pixels) of about 4,000. Compare with the histograms of TM bands 1, 2 and 3.

User-defined stretch applied to D:\images\V4.img
Good results are obtained by setting the upper and lower breaks at 138 and 28, respectively. The road network is sharply defined; even some small roads can be detected. The field boundaries and their patterns, especially south of Lough Ramor, are very clear.

Water bodies are easily detected as well as the small river, and can be distinguished from the darker vegetation patches. There is no evidence of haze on the image that could hinder its legibility.

User-defined histogram of D:\images\V4.img
The histogram shows almost 9,000 pixels with a zero DN value (this is shown on the histogram by the tall white bar). The remaining pixels are distributed across the whole 0–255 DN output range, with a peak near 166. There are also about 1,500 pixels that have been set to 255.

D:\images\V5.img image file (Landsat TM band 5; bandwidth 1.55–1.75 μm)

Landsat TM band 5 is in the reflected infrared portion of the electromagnetic spectrum and may provide information on moisture content of soil and vegetation. Again, as for TM band 4, its longer wavelength compared to the visible portion of the electromagnetic spectrum makes it less vulnerable to scattering.

Unstretched image of D:\images\V5.img
The image is medium grey with little contrast. Field patterns are not as distinct as on the D:\images\V4.img image and road patterns and settlements cannot be determined. Water bodies retain a very dark tone, similar to D:\images\V4.img.

Unstretched histogram of D:\images\V5.img
The pixels range from DN 0 to DN 163 with a mean value of 70. The histogram shows a distribution very similar to the one for D:\images\V4.img – again we have two peaks in the distribution. The bulk of the pixels are within the DN range 50–96.

Linear automatic stretch of D:\images\V5.img
The image is now brighter. The water features are all clearly visible. The differences in the signatures from the fields and the other vegetated areas are enhanced but it is still difficult to detect any field pattern. The settlements remain undetected on this image.

However, this TM 5 image shows two pale signatures enclosed by a dark grey signature on the northern shore of Lough Ramor which go undetected on the TM 4 image.

Histogram of linear automatic stretch applied to D:\images\V5.img
Most of the pixels are concentrated between DN 100 and DN 192. The pixel classes are now spread across the whole DN range. If this histogram is compared to the linear-stretched histogram for D:\images\V4.img, it should be evident that the distance between the two main surface classes (water and vegetation), as measured from their means, is greater for the TM 4 image than for the TM 5 image. Thus the separation of these classes is better on TM 4.

User-defined stretch applied to D:\images\V5.img
Note that the minimum and maximum DN values of the unstretched histogram are 0 and 163, respectively. Good results are obtained by setting the upper break at 103 and the lower break at 25.

The image shows quite good contrast, better than the linear automatic stretch. The image resulting from the user-defined stretch clearly shows all the water features and differences in reflectance in the cultivated areas as well as in the areas on the northern shore of Lough Ramor. The settlements remain undetected.

User-defined histogram of D:\images\V5.img
The histogram shows a peak of almost 9,000 pixels as black (DN value of 0) and about 1,500 pixels have been set to a value of 255. The bulk of pixels lie between DN values 120 and 192.

D:\images\V7.img image file (Landsat TM band 7: bandwidth 2.08–2.35 μm)

TM band 7 is in the reflected infrared portion of the electromagnetic spectrum and has applications in geology because of the presence of clay absorption bands. It is also sensitive to moisture content.

Unstretched image of D:\images\V7.img
The image is quite dark and only the lakes remain distinctive, though part of the N3 road may just be discerned. There are bright patches scattered across the image, mainly within the southeast quadrant.

Unstretched histogram of D:\images\V7.img
The histogram shows that the pixel values range from 0 to 91. There are two different peaks: the smaller at about DN 10 and the main one at about DN 25. There are few pixels with DNs greater than 40.

Linear automatic stretch of D:\images\V7.img
The image is now much brighter. The lakes are distinctive, though the river is difficult to detect for most of its length. There are a large number of bright spots scattered throughout the image, similar to those seen on D:\images\V3.img, many of which represent small settlements. The field patterns are detectable but field boundaries are difficult to determine.

Histogram of linear automatic stretch applied to D:\images\V7.img
The pixel classes cover the entire DN range from 0 to 255.

The histogram maintains the two peaks, the main one at about DN 128. The main histogram is markedly asymmetric, with DN values gently tailing off at higher digital numbers. Most of the pixels are concentrated between DN 100 and 192.

User-defined stretch applied to D:\images\V7.img

A good contrast is obtained with 40 and 10 as upper and lower break limits. All the water features are clearly visible and the N3 road is detectable for much of its length. The scattered buildings retain a bright response.

Rectangular field patterns may be discerned but TM band 7 is not the most suitable for obtaining information about vegetation cover.

User-defined histogram of D:\images\V7.img

The histogram no longer has two peaks because the values that constituted the smaller one were all set to zero (black) *by the lower break limit. There is now a good spacing of all the pixel classes across the DN range, with a single peak of the distribution at about DN 100. There are also about 3,000 pixels set to 255 (white) by the upper break limit.*

If you have completed the above practical, you should now be familiar with moving around the **DRAGON** environment. You should now repeat this exercise using other datasets. In addition, use other stretches such as the histogram equalisation which preferentially stretches the most populated parts of the histogram or Gaussian which redistributes the DN bins such that the histogram has a normal or Gaussian distribution. To exit, press any key then click on X in the menu. When prompted 'Do you wish to exit DRAGON' type Y, then press RETURN.

Practical Exercise 2
Grey-level thresholding and area measurement

The objectives of this practical are:

- to find the most suitable TM band within the Virginia dataset for the delineation of water bodies;
- to measure the areal extent of the lakes.

An examination of individual single-band images showed that the images acquired at longer wavelengths (D:\images\V4.img; D:\images\V5.img; and D:\images\V7.img) are less affected by scattering than the shorter wavelength images (D:\images\V1.img; V2.img; and D:\images\V3.img). Refer back to the Virginia images and decide which one allows the best delineation of the water bodies (we suggest D:\images\V4.img).

Grey-Level Thresholding

Thresholding is a technique that allows us to segment the DN values of an input image into two classes: one for those pixels having DN values below an analyst-defined grey level and one for those pixels above this value. All the pixels whose DN values fall below the analyst-defined grey level in the input image will be displayed as 0 (black) in the output image. All the pixels with DN values above that threshold will be set to DN 255 (white).

This exercise will create an output image with only two different grey levels: white and black. The resulting image will allow an easier detection of water features and therefore the measurement of their areal extent.

From the previous practical, you should now be familiar with contrast-stretching operations and the technique of moving the cursor and selecting and clicking to perform different image file display operations.

Start DRAGON by double clicking the DRAGON icon on the screen or from MS-DOS mode as explained in practical 1.

- Click D from the DRAGON Main Menu to display an image file.
- Click GRA to display the image in tones of grey.
- Note that the Image File line displays the default directory D:\images\. Click on the blue triangle at the end of the line and select the D:\images\V4.img image file (or another one if desired).
- The Histogram Adjustment line displays by default the LINEAR AUTOMATIC stretch option. Click on the blue triangle at the end of the line and select the NONE option; click it to apply.
- Click RUN at the bottom left corner of the screen to display the image.

In order to perform the thresholding of an image you have to set a DN value that will allow you to separate the water bodies from the other features. Note that D:\images\V4.img was acquired with TM band 4 (near infrared). Electromagnetic radiation in the near infrared is strongly absorbed by water and therefore the reflectance values of the pixels corresponding to water features will be low.

Display the histogram of D:\images\V4.img and examine it.

- Press ESC to exit the screen.
- Click R (Return to previous menu).
- Click U (Utility operations).
- Click HIS (Histogram of the image).
- The Image File line will display =M, the D:\images\V4.img file is still stored and the Apply Stretch displays NO by default. Leave unchanged.
- Click RUN on the bottom left corner of the screen.

Examine the histogram and establish what DN values correspond to the water features of the image.

Note that the histogram is characterised by a bimodal distribution of the pixels. There are two distinct peaks in the values, one around 6 and the other near 112. It is probable that the peak around 6 is caused by the signature for water, so that setting a DN value corresponding to the right of the distribution should allow us to separate water features from the rest of the image.

We found that a DN threshold value of 35 gives a good result, showing all the water features in the area of the image. You have to set all the values below the DN threshold value to 0, and those above it to 255. Therefore, you have to apply a stretch option that allows you to enter break limits.

- Press ESC.
- Click R.
- Click D.
- Click GRA.
- The Image File line still displays =M; press RETURN to make the Histogram Adjustment line active.
- Click on the blue triangle at the end of the line and select USER-DEFINED, click it and press RETURN.

You are now required to enter upper and lower break limits. Note that this time you have to enter the same value for both the UPPER and LOWER break limits. This will set all the pixels with a DN value above the threshold value to 255 (white) and the pixels with a DN value lower than the threshold to 0 (black).

Enter the DN threshold value of 35 and click on OK and then RUN.

The image displays water features in black on a completely white background. No other feature is displayed.

If you examined the histogram for this image you would find there are only two classes of pixels, one with 0 DN with few pixels and the class with DN 255 with a very high number of pixels.

Area measurement of Lough Ramor

- Press ESC.
- Click R to return to the Main Menu.
- Click G (Geometry operation).

- Click MEA (Measure lengths and areas). The Image File line still displays =M (D:\images \V4.img stored in memory).
- Press RETURN.
- Type METRES on the line 'Distance Unit' (first cancel the default unit using the backspace key), then press RETURN.
- Type 30 for the X scale factor because a TM Landsat pixel is 30 m across (first cancel the default value using the backspace key), press RETURN.
- Type 30 for the Y scale factor (first cancel the default value using the backspace key), press RETURN.
- Do not change the drawing colour option number.
- Click RUN.

You should now see the thresholded image displayed on the screen. Note that the mouse no longer has the shape of a small dragon but it is a cross. Also note that the upper line of the screen displays the co-ordinates of the cursor as line and pixels. The lower line of the screen shows different command options.

You are required to trace the contours of Lough Ramor (see Figure M2):

- Move the cursor to the bottom of the screen to the edge of the lake and click the left button of the mouse to start tracing and click it again any time you change direction to follow the contour of the lake.
- When you have traced the whole perimeter of the lake, returning to near the point you started from, press the F7 key to close the polygon.
- Press the F8 key to measure the area extent of Lough Ramor.

You should obtain an area of around 6,000,000 square metres or 6 square kilometres depending on how accurately you trace the outline.

If you wish, you can repeat the exercise for one of the smaller lakes on the upper left corner of the image. You may need to zoom in order to trace the outline of the lake more accurately. Position the cross-hairs on one of the lake margins (do NOT press the mouse button) and press Z to zoom in (this may be repeated). To unzoom press D. You can also find out the total area extent of the water features in the image by noting the results for the separate lakes and then summing them together. When you have completed the practical, press ESC to return to the Geometry menu, from which you may exit.

Practical Exercise 3
Investigation of the signatures of various surface types using multispectral datasets

An examination of the Virginia image files in the single-band practical allowed the investigation of individual images acquired at different wavelengths. However, those images contained data from a single band only. Combining data from a number of bands into a colour image allows more information to be extracted. Since the human eye can see only combinations of three colours, red, green and blue, all the images must be displayed in those colours irrespective of the wavelength at which the image was obtained. It is possible to display images take at wavelengths not visible to the human eye, provided that they are displayed in one of the above three colours.

The multiband datasets obtained by remote sensing platforms enable a large number of false colour combinations to be produced. This practical considers the formation of false colour composites for a number of datasets. The reader should be familiar with the formation of false colour composites.

The aims of this practical are:

- to provide hands-on experience in the production of false colour composites;
- to understand how choosing different band combinations can yield information on different surfaces;
- to investigate how surfaces appear different in the visible, infrared and microwave ranges of the electromagnetic spectrum;
- to produce a synergistic display using radar and TM data.

After displaying each image you are asked to describe it. A short description is provided in italics but you are encouraged to work through the exercises carefully, making appropriate notes as you progress. The duration of this practical is approximately four hours.

Probably the simplest colour image that one can produce using Landsat TM data is one displaying TM band 1 (visible blue) as blue, TM band 2 (visible green) as green and finally TM band 3 (visible red) as red. This results in an approximately natural colour image.

Before continuing, it is very important that you should know how to display images and change the stretch that is applied to them. If you are unsure how to accomplish this, you should redo the single-band contrast-stretching exercise (practical 1). The quality of your monitor can greatly affect the appearance of colour images. Excellent colour definition was given by a VESA Super VGA and the colours described in this and subsequent practicals refer to the colours observed on such a monitor.

Now create a colour image using the following three TM bands from the Northwest Ireland dataset.

D:\images\Don1.img	as blue
D:\images\Don2.img	as green
D:\images\Don3.img	as red

- It is assumed that you are at the Main Menu.
- Click D (Display images) from the DRAGON Main Menu.
- Click 3BA (3-band composite image display) from the Display menu. You are now asked to enter the name of the image file to be displayed as blue (the default directory D:\images\ is shown so just enter Don1.img). Click the blue triangle and select the image file you want, click it and press RETURN to make the next line of the screen active.
- You are required to select a histogram adjustment for the blue image file. LINEAR AUTOMATIC stretch option is displayed as default. Leave unchanged and press RETURN to make the next line of the screen active.
- Select the image file you want to be displayed as green (Don2.img) and press RETURN.
- You are required to select a histogram adjustment for the green image file. LINEAR AUTOMATIC stretch option is displayed as default. Leave

unchanged and press RETURN to make the next line of the screen active.

- Click on RUN at the bottom left of the screen to display the colour image on the screen.

Describe the image.

The colours observed on this image approximate to what the human visual system would detect if you were looking down on the area from the satellite.

The image shows several structural lineaments which represent faults (see also the filtering practical). Water bodies are shown in black and there are a number of small lakes in this region.

There is no evidence for extensive managed agriculture in this mountainous area and little sign of human habitation, but near the bottom left corner there is a blue-grey area which is the town of Glenties (see Figure M1). The attention is caught by the white and blue patches near the right margin of the image, which are caused mainly by the presence of clouds.

The predominant colour for vegetation in this scene is a red/brown signature that is due to heather and gorse which had died back when this winter image was obtained. Coniferous forest plantations are shown in a dark green colour. Small patches of grey which are scattered throughout the image are possibly regions of rock exposed at the surface.

A common false colour composite image is created using TM bands 2, 3 and 4. These three bands are usually displayed in blue, green and red, respectively. The resulting false colour composite will display healthy living vegetation in shades of red.

Display, using a histogram equalisation stretch for all bands, a false colour composite image using the following image files from the Northwest Ireland dataset:

D:\images\Don2.img   displayed as blue
D:\images\Don3.img   displayed as green
D:\images\Don4.img   displayed as red

Press ESC to return to the Display menu and then choose 3BA. You will need to delete =B before inputting the full file name D:\images\Don2.img. Then repeat for the other two bands.

Click on RUN at the bottom left of the screen to display the false colour composite on the screen.

Describe the image.

Forests are clearly defined on this image by their bright red colour. The reflectance of coniferous trees in this image has a high value compared to the poor reflectance of the other features present in this winter image but this may not be the situation in every season. It would be very useful to compare two images of this area acquired at different times of the year to examine the differences in colours in which these coniferous trees would be rendered.

The area seems to be divided into two different zones: the upper left part of the image is mainly a grey colour whereas a red tinge predominates in the southeast. This is partly due to topographic variations but the areas are underlain by different rock types. One would therefore expect differences in soil, porosity and permeability which could then influence the type and degree of vegetation cover.

Create another false colour composite, using the instructions given earlier, with the following image files from the Northwest Ireland database:

D:\images\Don5.img   (IR) displayed as blue
D:\images\Don4.img   (IR) displayed as green
D:\images\Don7.img   (IR) displayed as red

The human visual system is unable to 'see' at these wavelengths. Click on RUN at the bottom left of the screen to display the false colour composite on the screen.

This combination, using long-wavelength bands, produces quite a sharp image. Forests are shown as green because of the relatively high reflectance in TM 4. The presence of vegetation along the valleys is also displayed in green. Two red signatures, caused by high reflectance in TM 7, are very prominent in the northwest corner.

Create, using a histogram equalisation stretch for all

bands, a false colour composite using the following image files from the East Ireland dataset:

D:\images\May3.img (red) displayed as blue
D:\images\May5.img (IR) displayed as green
D:\images\May4.img (IR) displayed as red

Click on RUN at the bottom left of the screen to display the false colour composite on the screen.

This false colour composite image is different from the Northwest Ireland images because it clearly shows cultivated fields mostly with a characteristic red/orange and olive green signature. Most of the former colours represent pasture whereas the latter signature is generally associated with late crops. If these fields were examined one month later, they would be associated with different colours. Large pale blue areas represent bare soil while some of the dark red/brown areas represent forests. A single field of oil seed rape with a very distinctive pink colour is seen south of centre. The image shows three small towns (Maynooth, Leixlip and Celbridge, see Figure B3) with a characteristic blue signature. The road between Leixlip and Maynooth is clearly defined. Along this road, close to Leixlip, an isolated blue signature for an industrial estate can be discerned. This is the Intel plant, where many of the world's computer chips are manufactured (possibly even the one in the computer that you are using). The River Liffey between Celbridge and Leixlip is clearly defined by a black signature. One of the most distinctive features on the image is an approximately elliptical shape that occurs just north of centre. This is totally unlike the rest of the scene and shows no evidence of the interlocking field patterns that characterise the region around it. This is Carton estate, which has remained virtually unchanged over 250 years.

A false colour composite can also be created by using three bands from a sensor other than Landsat TM. The East Ireland dataset contains three radar image files obtained in 1994 aboard the Space Shuttle.

D:\images\Mayr1.img L band (wavelength 23.5 cm)
    HH polarisation
D:\images\Mayr2.img L band (wavelength 23.5 cm)
    VV polarisation

D:\images\Mayr3.img C band (wavelength 5.8 cm)
    VV polarisation

Before beginning this part of the practical, the reader should be totally familiar with the parameters that control a radar image. If necessary, consult section 3.2 of *Introductory Remote Sensing: Principles and Concepts*.

Briefly, radar images show the intensity of the returns from ground features. The intensity of the return is related to the geometric and electric characteristics of the terrain. One of the most important characteristics affecting the intensity of radar return signals is the surface 'roughness' of ground features. Roughness is related to the wavelength of the incident microwave: a surface that is rough at one wavelength may be smooth at another. A radar-rough signature appears bright and a radar-smooth surface appears dark on a radar image.

Use the following band combinations (LINEAR AUTOMATIC) to produce a colour radar image.

D:\images\Mayr1.img L band (wavelength 23.5 cm)
    HH polarisation displayed in blue
D:\images\Mayr2.img L band (wavelength 23.5 cm)
    VV polarisation displayed in green
D:\images\Mayr3.img C band (wavelength 5.8 cm)
    VV polarisation displayed in red.

Describe the false colour composite image.

This false colour composite image is composed of shades of red/orange, pale lilac and green. Note the grainy appearance of the radar image compared to the Landsat TM image of the same area.

The small towns are shown in white, indicating high radar returns due to the presence of corner reflectors in urban areas. A bright signature extends around the northern and eastern perimeter of Carton estate because of the high degree of scattering from the trees in these areas.

The water body at the lower right corner of the image is quite dark because water acts as a radar-smooth surface and reflects the microwave energy away from the sensor.

Fields with a soil surface appear a very pale orange in

this image whereas those with late crops are generally green, while pasture and early crops are shown in red.

Note the presence of an additional linear feature in this image compared to the TM false colour composite. The radar image was taken a number of years after the TM image and a new E–W-trending road to the south of Maynooth and Leixlip can just be detected. For this scene, because it is dominated by surfaces which have different signatures in the visible and near infrared rather than marked differences in the microwave range, TM is the better system. However, in other situations radar often proves significantly better. For example, a TM image of the area covered by the Sudan dataset would show very little because of the extensive sand cover.

We have seen the advantages of using data from three wavelengths and merging them into one single image. It is also possible to merge information from different sources. For example, two TM bands and one radar band can be combined into a single false colour composite. (See section 3.8, where synergistic displays are discussed).

From the East Ireland dataset select the following files and create a false colour composite image:

D:\images\Mayr1.img	displayed as blue
D:\images\May3.img	displayed as green
D:\images\May4.img	displayed as red

This false colour composite image will display information from two Landsat TM bands (TM 3, visible red, and TM 4, near IR) and information from the radar L band with HH polarisation.

- Select the image file you want to be displayed as blue (Mayr1.img) and press RETURN.
- Select the USER-DEFINED contrast stretch option and enter 128 and 20 as upper and lower limits for the image file to be displayed as green and click OK.

- Select the image file you want to be displayed as green (May3.img) and press RETURN.
- Select the USER-DEFINED contrast stretch option and enter 32 and 18 as upper and lower limits for the image file to be displayed as green and click OK.
- Select the image file you want to be displayed as red (May4.img) and press RETURN.
- Select the USER-DEFINED contrast stretch option and enter 150 and 90 as upper and lower limits for the file to be displayed as red; click OK and then RUN.

Describe the image.

The false colour composite shows two markedly different areas in the image: the area where only the two TM bands are displayed and the central region where the two TM bands overlap with the radar L-band image.

The central part of the image shows the three settlements displayed as bright cyan signatures due to the presence of corner reflectors within the built-up areas. There is a possibility that this could be confused with areas of soil. The pasture in Carton stands out as a prominent red colour and the forested perimeter is a dark blue.

Vegetation in the fields tends to be displayed in shades of red and purple.

Note the presence of a very bright yellow signature in the middle lower part of the false colour composite image which is oil seed rape.

You should now experiment with different combinations of radar and TM bands for a range of user-defined contrast stretches. The spectral signatures of other features should also be investigated by means of the other datasets.

To finish the session, press ESC to take you back to the Display menu and then click X to exit DRAGON. You will be prompted (Y/N)? Type Y and then press RETURN.

Practical Exercise 4
Ratio imagery for the production of vegetation indices and geological mapping

Band ratioing is the enhancement process of dividing the DN values of the pixels of a spectral band by the corresponding DN values of pixels of another band (see section 3.4 for further details). This technique has two advantages: it reduces the influence of topographic effects on the imagery and it emphasises the differences between the reflectance curves of different substances, showing the gradient change between bands. These advantages have made ratioing a widely used technique in vegetation and geological studies.

The objectives of this practical are:

- to perform simple arithmetic procedures on images (addition, subtraction and division) and to save the results to a separate file;
- to investigate the usefulness of ratio imagery;
- to compare ratio images with standard false colour composites for geological investigations.

Ratio images can be used to generate false colour composites by combining three monochromatic ratio datasets and displaying them in blue green and red. Ratio images can be used to estimate the vegetation density of an area. A common ratio used for vegetation studies is the TM 4/TM 3 ratio or Simple Vegetation Index. This ratio exploits the fact that visible red light (TM 3) is absorbed by vegetation whereas near-infrared radiation (TM 4) is strongly reflected by vegetation. This practical should take approximately four hours. Images from the Northwest Ireland dataset are used for vegetation indices and images from the Andes dataset for the geological investigation.

Perform a ratio of TM 4 on TM 3 band using the image files from the Northwest Ireland dataset. Enter the DRAGON environment, then:

- Click E (Enhance images) from the Main Menu.
- Click RAT (Ratio of two image bands).

- Select the file D:\images\Don4.img for Image File 1 by clicking the blue triangle and clicking the file name.
- Select the file D:\images\Don3.img for Image File 2 by clicking the blue triangle and clicking the file name.
- Retain the grey display option.
- Select LINEAR AUTOMATIC from the histogram adjustment option.
- Do not change any of the other options.
- Click RUN at the bottom left corner of the screen to display the ratioed image.

Describe the image.

Pale areas on this ratio image show where the DN value is greater in TM 4 than in TM 3. The region is generally characterised by a mid-grey signature within which some variation can be detected. The southern part of the image is slightly paler than the upper part. The most dominant features on the image are the forests, which are displayed as white. A ratio image such as this is ideal for measuring the area of the forests because the boundaries are sharply delineated.

The low reflectance of water in the infrared means that it is shown as black in this ratio image. Thus the Gweebarra River and Lough Muck (near the right margin of the image) are clearly defined. Note that this image shows almost no evidence of the topography of the area.

There are also more complex ratios that involve differences and sums between spectral bands. A commonly used one is the Normalised Difference Vegetation Index (NDVI), defined as:

$$(\text{near IR} - \text{red}) / (\text{near IR} + \text{red})$$

To display the NDVI image using TM bands 3 and 4, it is necessary to obtain the difference between the two image bands, save the resulting image, then obtain the sum of the same two image bands and save the result. Finally, the two saved images are ratioed.

- Press ESC to go back to the Enhancement menu.
- Click DIF (Difference of two image bands).
- Select file D:\images\Don4.img for the Image File

1 line by deleting =M and inputting the full file name (including paths).

- Select the file D:\images\Don3.img for the Image File 2 line by clicking the blue triangle and clicking the file name.
- Retain the grey display option.
- Perform a LINEAR AUTOMATIC stretch from histogram adjustment.
- Click RUN at the bottom left corner of the screen to display the difference between the two image bands.

Save the image to a floppy disk:

- Press ESC twice to go back to the DRAGON Main Menu.
- Click F (File Handling operations).
- Click SAV from the File Handling menu.
- If you wish to save it to a floppy disk, delete the current line and type: A:\DIFF43.img (or another name). If you wish to save it to your hard disk, ensure that you type in the entire path, e.g. c:\dragon\images\diff43.img would save the file in the subdirectory images which is within the DRAGON directory.
- Click RUN at the bottom left of the screen.

Now you have to perform the sum of the two image files D:\images\Don4.img and D:\images\Don3.img.

- Press ESC to go back to the DRAGON Main Menu.
- Click E (Enhance images).
- Click SUM (Sum of two image bands).
- Select D:\images\Don4.img for Image File 1 after deleting = M.
- Select D:\images\Don3.img for Image File 2.
- Choose a grey display option and a LINEAR AUTOMATIC Stretch.
- Leave the scaling option as the default (average).
- Click RUN at the bottom left corner of the screen to display the difference between the two image bands.

Save the image to a floppy disk as described above but name it A:\SUM43.img. Now you can create an

image of the NDVI of TM bands 3 and 4 of the Northwest Ireland dataset.

- Press ESC to go back to the DRAGON Main Menu.
- Click E (Enhance images).
- Click RAT (Ratio of two image bands).
- Remove =M and select A: \DIFF43.img (or from hard disk if saved there) for Image File 1.
- Select A: \SUM43.img (or from hard disk if saved there) for Image File 2.
- Select LINEAR AUTOMATIC from the histogram adjustment option.
- Click RUN at the bottom left corner of the screen to display the Normalised Difference Vegetation Index of the two image bands.

Describe the image.

The NDVI image retains a substantial amount of the topographic variation within the area and allows many of the lineaments to be detected. Water bodies are sharply defined in bright tones. However, not all bright signatures correlate with water features. Suitably orientated slopes are also shown bright in this image. Note that on this image forests are suppressed. They appear as a dark grey which shows little contrast with the surrounding terrain. Thus, in this case, the NDVI has not accentuated the forests as the simple ratio image did.

Exploration for a Geological Ore Body Using Ratio Imagery

As mentioned previously, it is possible to create three different monochromatic ratios and then display them as blue, green and red to generate a false colour composite ratio image. Such images often have applications in the search for ore bodies. We wish now to create a false colour composite with three ratio images using the TM image files from the Andes dataset. You must create three ratio images, save them and then display them in red, green and blue to create a false colour composite.

Before you create the colour ratio image, you should first examine and describe various false colour composites produced by using the individual bands in the Andes dataset. Suggested combinations to examine (using the instructions given in the multi-spectral practical) are:

D:\images\Peru1.img   as blue
D:\images\Peru2.img   as green
D:\images\Peru3.img   as red

D:\images\Peru3.img   as blue
D:\images\Peru5.img   as green
D:\images\Peru4.img   as red

D:\images\Peru7.img   as blue
D:\images\Peru5.img   as green
D:\images\Peru4.img   as red

When you have examined and described those images you should:

- create a ratio of D:\images\Peru1.img /D:\images\Peru2.img and save the image to a file called C:\dragon\images\ratio12.img using the instructions given earlier;
- create a ratio of D:\images\Peru3.img /D:\images\Peru4.img and save the image to a file called C:\dragon\images\ratio34.img using the instructions given earlier;
- create a ratio of D:\images\Peru5.img /D:\images\Peru7.img and save the image to a file called C:\dragon\images\ratio57.img using the instructions given earlier.

Now display the three ratio images that you created in a false colour composite image. Display the files as:

C:\dragon\images\ratio12.img   as blue
C:\dragon\images\ratio34.img   as green
C:\dragon\images\ratio57.img   as red

To achieve this:

- Press ESC until you access the DRAGON Main Menu.
- Click D.

- Click 3BA.
- Type C:\dragon\images\ratio12.img in the Image File for Blue line after deleting =B.
- Type C:\dragon\images\ratio34.img in the Image File for Green line after deleting =G.
- Type C:\dragon\images\ratio57.img in the Image File for Red line after deleting =R.
- Apply an equalization stretch to all the bands. Click RUN at the bottom left of the screen and display the false colour composite image.

Describe the image that appears on the screen.

The image is much more colourful than the standard false colour composite and is dominated by a green signature except for a bright pink/purple colour in the top right. Note the presence of an anomalous yellow/orange elliptical area in the upper left corner of the image which shows the location of a copper ore body. This region of mineralisation is not as apparent on the false colour composites produced by using the individual TM bands. Similar signatures can be detected in the southeast quadrant of the image which also represent regions where copper ores have been found.

Using information learned in the Threshold practical you can measure the area of the ore body.

- Press ESC.
- Click R to go to the Main Menu.
- Click G (Geometry operation).
- Click MEA (Measure lengths and areas). The Image File line displays =M. Change this to =C in order to redisplay the colour ratio image.
- Press RETURN.
- Type METRES on the line Distance Unit (cancel the default unit using the backspace key). Press RETURN.
- Type 30 for the X scale factor (cancel the default unit using the backspace key). Press RETURN.
- Type 30 for the Y scale factor (cancel the default unit using the backspace key). Press RETURN.
- Do not alter the drawing colour or data output line.
- Click RUN.

Measure the area extent of the ore body in the upper left of the image and take note of your result.

- Move the cursor to the top left of the screen and start tracing the ore body.
- Click the left button of the mouse to start tracing and click it again any time you change direction to follow the outline of the ore body.
- When you have traced the whole perimeter of the ore body, returning to the point you started from, press the F7 key to close the polygon.
- Press the F8 key to measure the area extent of the ore body.

You should obtain an area of around 2,950,000 square metres though your answer may vary somewhat.

It is possible to create 455 different colour ratio images using a single six-band dataset. Experiment using different ratio combinations on the following datasets: Virginia, Northwest Ireland, Peru and Nevada and compare the colour ratio images with false colour images produced by using individual bands. You may also produce ratio images incorporating both TM and radar data by using the East Ireland dataset.

Practical Exercise 5
Convolution filtering

Filtering was introduced in section 3.5 and the reader should ensure that he/she is familiar with the concepts and terminology of that section.

In summary, an image can be thought of as being formed of sinusoidal waves of different wavelengths. A low-frequency part of an image represents an area where the DNs change gradually, while a high-frequency part is where there are sharp variations in DN. The high-frequency components, or sharp tonal changes, represent the local details whereas the low-frequency components, or gradual changes in tone, represent large-scale or regional patterns. These high- or low-frequency components can be enhanced or subdued by the use of filters. High-pass filters are ones that emphasise the high-frequency components and are useful for accentuating edges. Conversely, low-pass filters let the low-frequency components of an image pass and hence smooth the image.

DRAGON allows simple convolution filtering in the spatial domain of remotely sensed datasets. Before starting to filter the images, you will be required to display them and take notes on their appearance. **If you are unsure how to display images, you should revise the instructions given in practical exercise 1.** The aim of this exercise is to investigate the effects of various filters upon the appearance of an image. Two image files will be used to illustrate this practical: D:\images\V4.img from the Virginia dataset and D:\images\Peru4.img from the Andes dataset. This practical should take approximately four hours.

You should now display, using a linear automatic stretch, the following image file: D:\images\V4.img.

Describe the image.

The image is quite bright. There are several lakes in the area, the largest (Lough Ramor) in the lower part of the image. Two other lakes are prominent in the northwest quadrant. A river is seen to flow between a small lake and Lough Ramor. The road network can also be determined, especially the main N3 road which runs from the upper left corner to the bottom right. The N3 road passes through the town of Virginia (see Figure M2). Isolated settlements are just visible as small grey dots. Apart from the water bodies, the most striking feature of this image is the field pattern: the whole area displays many small fields in different tones of grey. There are also darker vegetated areas in the bottom right portion of the image and on the northern shore of Lough Ramor.

To apply a filter to the image:

BOX C

- Press ESC to exit the image.
- Click R using the mouse (return to the previous screen) to access the DRAGON Main Menu.
- From the DRAGON Main Menu move the cursor onto E (enhance images) and click with the left button of the mouse.
- Move the cursor on the line FIL (filtering applied to an image) and click it.
- Note that the first line of the screen, Image File, will display the prompt =M (the image file last displayed is stored in the computer memory). Leave unchanged.
- Note that the second line of the screen, Kernel type (or Filter type), displays the Smoothing filter option by default. Leave unchanged and press the RETURN key.
- The active line is now Display option. Leave the default option Grey unchanged and press the RETURN key.
- The Histogram Adjustment line is now active. It displays NONE by default. Click the blue triangle and select LINEAR AUTOMATIC by clicking on it.
- Move the cursor to the bottom left of the screen and click RUN.
- Note that after a few seconds a blue line appears at the top margin of the screen, stating how many pixels were truncated, and informs you that to continue you need to press a key. Do so. (Depending on the particular filter that is used, values greater

---

than 255 or less than 0 may be calculated. These are truncated to 255 and 0 respectively. More sophisticated image processing software may allow the display of 32-bit real numbers).

Now describe the result.

The image appears out of focus, the lakes are visible but the water–land interface is more diffuse. The major roads are still visible as well as the small town of Virginia. It is much more difficult to have a precise perception of the fields because their boundaries are blurred.

Note: Some stretches result in a dark image which may be difficult to interpret. However, these images may be contrast stretched for easier viewing. Also one may wish to save the results of a particular stretch. The SAVe option in the File Handling menu saves the last displayed image to a file which can be displayed at a later date. Refer to the ratio imagery practical for instructions on saving an image to file.

The edge filter used in DRAGON is

-1	-1	-1
-1	$+9$	-1
-1	-1	-1

To apply the EDGE filter you must first reload D:\images\V4.img.

- Press ESC twice to go back to the DRAGON Main Menu.
- Redisplay D:\images\V4.img, again using an equalisation stretch, and then follow the instructions in BOX C but this time apply the EDGE filter for Kernel type.

The road network is well displayed, including some of the minor ones. The field patterns are extremely well defined and it is possible to distinguish even very small fields because their boundaries are now sharp. The small islands in Lough

Ramor are also observed better though the town of Virginia is more poorly contrasted against the background and is less easy to detect.

The Filtering option in DRAGON allows you to design your own 3×3 filters in order to enhance or suppress particular features in the image by using the USER-DEFINED kernel. The user-defined option allows you to create a 3×3 filter. Note that you are requested to enter the numbers in the kernel by column. Thus to apply the following 3×3 filter:

1	2	3
4	5	6
7	8	9

you need to input 1 4 7 2 5 8 3 6 9, separating the numbers by a space.

Now try the following filter: $-1 -1 0 -1 0 1 0 1 1$ (by entering the numbers in this order and separating them by spaces).

BOX D

- Display D:\images\V4.img using an equalisation stretch. Now press ESC, then, using the mouse, click R on the displayed menu.
- Click E (Enhance images) from the Main Menu.
- Click FIL; the Image File line displays =M as before but now it is the unprocessed image file that is stored in the computer memory; press RETURN.
- Click the blue triangle at the end of the Kernel type line and select USER-DEFINED; click it.
- Press RETURN. Note that the Kernel values by column is active.
- Delete the values using the BACKSPACE key and Type in the new filter values shown above.
- Click on the histogram adjustment line and select LINEAR AUTOMATIC. Click it.
- Click RUN at the bottom left of the screen.

Note that the resulting image is very dark and tends to highlight the edges of some of the features. You need to add this image to the original one to produce an image that is easier to interpret.

BOX E

- Press ESC to exit the image and access the Enhancement menu.
- Click SUM (Sum two image bands).
- Note that the Image file 1 displays =M (the image resulting from the filtering is stored in the computer memory), and press RETURN.
- Now the Image file 2 is active and displays the default directory D:\images\.
- Select the Image file D:\images\V4.img again to sum the two images.
- Click on the Histogram adjustment line and select equalisation. Click to apply.
- Do not change any of the other options.
- Click RUN at the bottom left corner of the screen.
- (This image can be saved by using the SAVe option in the File handling menu.)

Describe the image.

The resulting image is now brighter and has a slight 3D appearance. The road network, the small river, the lakes and the boundaries of the fields seem engraved as do some of the darker patches of vegetation. The image also looks as if it is illuminated from the southeast.

Try the following filter: 0 0 −1 0 0 0 1 0 0 (this is another filter used to enhance edges). To achieve this, reload D:\images\V4.img (stretched), go into the Filter option, pick USER-DEFINED for the Kernel type and enter the above numbers.

Note that the resulting image is very dark and shows only some of the edges of the features. You need to add

this image to the original one to have an image that is easier to interpret.

Add the filtered image back to the original one using the instructions in BOX E.

Describe the resulting image.

Again, this directional filter produces a slight 3D effect in which the edges of some features stand out from the background. This kernel appears to have preserved even the narrowest lineaments. The road network is clearly visible as well as the small river and many of the field boundaries.

Apply the following user-defined LAPLACIAN filter: 0 −1 0 −1 5 −1 0 −1 0 (note that the numbers must be separated by a space) using the instructions in BOX D.

Describe the image after you applied the filter.

The image is much sharper than the original unfiltered image and features are displayed in greater detail. The road network and field boundaries are very clearly defined and even minor roads can be observed. The small river is well defined and can be observed continuing to the very small lake in the very top right margin of the image. The result of this kernel is an image very similar to the one resulting from the EDGE in-built option. The only difference is that this kernel produces a slightly darker image.

It seems that only the EDGE filter option and the LAPLACIAN kernel produced a better image to interpret with most features enhanced with respect to the image produced using only the contrast-stretching option. The comparatively poor results for the other filters is possibly due to the lack of a predominant trend in the lineaments of the D:\images\V4.img image apart from the major roads, and the small river, field boundaries and other features have their edges aligned in different directions.

Design your own kernel to filter the image file D:\images\V4.img, following the above examples. Jensen (1996) provides many different examples of filters. Compare the results of your own kernels with those resulting from the above examples. From the application of the filters suggested by the above examples you should have noted that some kernels enhance edges running in one particular direction. This characteristic might be useful when filtering images which have lineaments with a predominant trend.

Display and describe the image D:\images\Peru4.img file using an equalisation stretch.

This image is totally unlike the Virginia dataset. It is very mountainous and there is no evidence of cultivation. Two rivers running in deeply incised valleys flow in a southerly direction and converge. The image has a general mid-grey tone, though a paler elliptical body may be observed in the northwest. There are a number of lineaments within the image with different trends. They tend to be accentuated because of the contrasting illumination in their shadowed and sunlit slopes.

Apply an edge-enhancement filter as explained previously and describe the effect of the EDGE filter on the image.

The image resulting from the application of the EDGE filter has less contrast than the original image, with less contrast across the lineaments. However, as expected, there is more detail in the topography. Fine lineaments with an ENE–WSW trend can be seen in the lower third of the image. This trend is not present in the upper part of the image.

You are now required to use a USER-DEFINED kernel, as described in BOX D. Try the following kernel: 1 −1 −1 1 1 −2 −1 1 1 1 (enter the numbers in this order).

The resulting image is characterised by NE–SW-trending bright edges. The rivers can still be detected. You need to add this image to the original one to have an image that is easier to interpret.

Add the filtered image back to the original one using the instructions in BOX E.

Describe the image.

The image shows the ridges facing northeast greatly enhanced with an apparent pseudo-illumination from the northeast giving a 3D effect.

Now try the following kernel: 1 1 1 −1 −2 1 −1 −1 1 (enter the numbers in this order).

Note that the resulting image is dark and shows only some of the edges of the features. However, there does appear to be a distinct dendritic pattern to many of the lineaments, possibly reflecting a fine drainage network. This is supported by the observation that these lineaments are associated with the two main rivers. You need to add this image to the original one to have an image that is easier to interpret.

Add the filtered image to the original one using the instructions in BOX E.

The image now appears as if it is a plateau incised by valleys. The northeast illumination-like effect gives the idea that the slopes facing northeast are flat areas, especially in the lower portion of the image. The main rivers do not now appear to be in valleys. The dendritic pattern is now also lost.

Now try the following kernel: 1 0 0 0 0 0 0 0 −1 (enter the numbers in this order).

The image again shows the lineaments with a bright signature. A WNW–ESE trend may be discerned in the lower half of the image. You need to add this image to the original one to have an image that is easier to interpret.

Add the filtered image back to the original one using the instructions in BOX E.

The results of the application of this kernel are quite good. The image shows a great deal of detail in a range of grey tones. The topography is well rendered and is quite similar to the original unfiltered image but sharper.

Apply the user-defined LAPLACIAN filter 0 −1 0 −1 5 −1 0 −1 0 (note that the numbers must be separated by spaces). Describe the effect of the filter on the image.

The image is quite bright. The difference in texture between the upper and the lower part of the image (where ENE–WSW fine lineaments can be discerned) is clearly detectable following the application of this filter. Tonal variations, quite similar to the original image, are maintained to some extent.

You should now experiment with different filters and apply them to various images that are included within the datasets.

Practical Exercise 6
Principal components analysis

Before beginning this practical, it is important that you:

1 know how to display single-band back and white and colour imagery which has been discussed in previous practicals;
2 understand the concepts discussed in section 3.7, on principal components.

An advantage of using principal components is that most of the variance in a multispectral dataset can be displayed using the first three principal components. Also the zero correlation between bands means that the resultant colour image is saturated with brilliant colours. The aims of this practical are:

● to compare and contrast principal component images;
● to examine the statistics of principal components;
● to examine various false colour combinations of principal component images;
● to produce scatterplots of image data in order to assess qualitatively the correlation between bands.

This practical should take approximately six hours. It is not possible to create principal component images with the version of DRAGON included with this book. However, principal component images have been produced for the following datasets in order that the effects of the principal components transform can be investigated and compared with images produced from the original data:

Northwest Ireland
Andes
East Ireland
Nevada
Virginia

These principal component images can be viewed either individually using the GRA option or as three-band colour composites (3BA option). The images have been processed so that they do not require any stretching and thus use the default option for histogram adjustment.

This exercise will use the Northwest Ireland dataset.

Northwest Ireland dataset statistics

Variance–covariance matrix

	Band 1	Band 2	Band 3	Band 4	Band 5	Band 7
Band 1	12.93	8.4	14.5	23.2	29.9	11.1
Band 2		7.5	12.8	22.4	29.1	10.8
Band 3			24.8	40.4	58.0	21.9
Band 4				100.3	117.5	40.5
Band 5					186.7	68.5
Band 7						27.3

Eigenvalues

	PC1	PC2	PC3	PC4	PC5	PC6
	322.6	20.9	11.7	2.3	1.4	0.5
% var.	89.7	5.8	3.3	0.65	0.4	0.15
Cum. % var.	89.7	95.5	98.8	99.45	99.85	100.0

Eigenvectors

	PC1	PC2	PC3	PC4	PC5	PC6
TM 1	0.14	0.2	0.66	0.69	0.13	0.16
TM 2	0.13	0.14	0.33	−0.16	−0.06	0.9
TM 3	0.25	0.08	0.56	−0.65	−0.25	0.38
TM 4	0.52	0.77	−0.35	−0.03	0.13	0.06
TM 5	0.75	−0.50	−0.148	0.20	−0.34	0.04
TM 7	0.27	−0.31	0.08	−0.21	0.88	0.004

The first three principal components contain 98.824 per cent of the total variance within the whole volume of data of the six bands (Landsat TM 1, 2, 3, 4, 5, 7). Thus, a false colour composite image created with the first three components will display virtually all the variance.

DRAGON software allows the production of scatterplots, which are two-dimensional representations of the DN value for two bands. The extent of the correlation between the bands can be qualitatively assessed by the shape of the distribution. The more elliptical the distribution, the greater the degree of correlation. Conversely, a scatterplot with an almost circular shape indicates that the two bands are uncorrelated and they provide independent information. The use of the scatterplots will be a useful support for the choice of the most suitable bands for the creation of false colour composite images.

To display the scatterplots of D:\images\Don5.img and D:\images\Don7.img

- Enter the DRAGON environment.
- Click U, Utility Operations from the DRAGON main menu.
- Click SCA Scatterplot of Image Bands.
- Input the name of the image file for the first band by clicking the blue triangle at the end of the line (or inputting manually D:\images\Don5.img) and press RETURN.
- Input the name of the image file for the second band by clicking the blue triangle at the end of the line (or inputting manually D:\images \Don7.img) and press RETURN.

- Leave the skip factor at 2, click RUN at the bottom left of the screen to display the scatterplot.

The scatterplot is elliptical in shape. White shows the highest concentration of pixels and blue the lowest. Try other band combinations from the Northwest Ireland dataset, for example D:\images\Don1.img and D:\images\Don.4img. There is less correlation for these bands as the distribution is less elliptical in character.

Examine the single image files of the six principal components of the Northwest Ireland dataset.

D:\images\Donpc1.img
D:\images\Donpc2.img
D:\images\Donpc3.img
D:\images\Donpc4.img
D:\images\Donpc5.img
D:\images\Donpc6.img

Display D:\images\Donpc1.img as a single-band grey image with no histogram adjustment. (Refer back to practical 1 if you are unsure how to do this).

From the eigenvalues table this image contains 89.7 per cent of the total variance of the dataset. The image is similar to a traditional image file. Note also the similarity of this band to TM 4 and especially to TM 5, D:\images\Don4.img and D:\images\Don5.img, respectively. This can be determined from the eigenvector matrix which shows that the first principal component is heavily weighted by TM 4 and TM 5. The image shows a good contrast and the forests yield a prominent mid-grey tone.

Repeat the single-band display for the other principal components and describe what you see.

D:\images\Donpc2.img
PC2 contains 5.8 per cent of the total information of the dataset. This image is quite different from D:\images\ Donpc1.img and from the original TM images. The clouds are particularly noticeable by the bright tone on the eastern edge of the image. Although many of the long lineaments present in this scene are not well displayed on this band, the mid-grey areas show a substantial amount of textural information.

D:\images\Donpc3.img
This image appears 'noisier' than the previous two but gross tonal variations are evident. The pale northwest region contrasts well with the mid-grey southeast part of the image. Forests again remain quite prominent.

D:\images\Donpc4.img
Principal component 4 information content is almost entirely made of noise. The image is very bright but only a few topographic elements can be determined. This image contains only 0.65 per cent of the variance of the dataset made of six bands. This image has no similarities with any of the original bands.

D:\images\Donpc5.img
Although this image contains only a very small proportion of the variance, 0.4 per cent, it still contains some elements of geographical information. Traces of some of the lineaments are discernible as dark tones.

D:\images\Donpc6.img
No useful information can be obtained from this principal component.

PC1 + PC2 + PC3 account for 98.8 per cent of the total variance. Hence with only three bands, or 50 per cent of the volume of the data, we can display almost the entire information content of within the dataset. To display these bands:

Display D:\images\Donpc1.img; D:\images\Donpc2. img; and D:\images\Donpc3.img as a false colour composite in blue, green and red respectively. (Refer back to practical 3 if you are unsure how to do this).

Describe the image.

The colours on this principal component image are very saturated and brilliant. However, the colour key is unusual: water features, for example, are no longer displayed as black, as in most conventional false colour composite images. The lakes in the area are brown/orange and well delineated. The colour differences are much greater than can be achieved on false colour composites that use the TM bands. The uplands in the northwest are characterised by a bright purple that is relatively uniform over a large region. This contrasts with the varied blue, yellows and green in the southeast. One can produce other principal component colour combinations. Experiment with some.

The Virginia, East Ireland and the Andes datasets also include six principal component image files which can be investigated and compared with individual TM bands. The Nevada dataset contains five principal components. You should also produce scatterplots for the individual bands. Use the principal component images of the Andes dataset to determine whether the copper ore deposit, which was evident on the ratio imagery (practical exercise 4) can be detected on the principal component imagery. If so, on what PC is it most obvious? Also, the principal components for the East Ireland dataset were calculated using TM and radar data so they should be examined to ascertain which image type contributed to which principal component.

Virginia, Ireland dataset statistics

Variance–covariance matrix

	Band 1	Band 2	Band 3	Band 4	Band 5	Band 7
Band 1	6.2	3.6	6.8	−10.2	14.3	9.9
Band 2		4.1	5.5	11.0	24.4	10.9
Band 3			12.8	−25.4	30.5	20.1
Band 4				780	300.3	42.7
Band 5					322.1	111.8
Band 7						52.2

Eigenvalues

	PC1	PC2	PC3	PC4	PC5	PC6
	937.8	225.1	8.97	3.29	1.2	0.72
% var.	79.67	19.1	0.76	0.28	0.10	0.09
Cum. % var.	79.67	98.77	99.53	99.81	99.91	100.0

Eigenvectors

	PC1	PC2	PC3	PC4	PC5	PC6
TM 1	−0.002	0.097	0.57	−0.45	0.67	0.14
TM 2	0.02	0.09	0.30	−0.32	−0.30	0.84
TM 3	−0.007	0.21	0.52	−0.122	−0.66	0.49
TM 4	0.89	−0.43	0.15	0.06	−0.026	0.019
TM 5	0.45	0.76	−0.38	−0.27	0.02	0.05
TM 7	0.10	0.42	0.39	0.78	0.17	0.17

Andes dataset statistics

Variance–covariance matrix

	Band 1	Band 2	Band 3	Band 4	Band 5	Band 7
Band 1	391	266	430	330	636	398
Band 2		193	320	256	514	317
Band 3			548	442	914	560
Band 4				402	805	482
Band 5					1,816	1,083
Band 7						666

Eigenvalues

	PC1	PC2	PC3	PC4	PC5	PC6
	3,751	213	29.8	11.4	10.3	1.02
% var.	93.4	5.3	0.74	0.28	0.26	0.02
Cum. % var.	93.4	98.7	99.44	99.72	99.98	100

Eigenvectors

	PC1	PC2	PC3	PC4	PC5	PC6
TM 1	0.27	0.7	−0.19	−0.39	−0.42	−0.27
TM 2	0.21	0.323	−0.03	0.005	0.11	0.92
TM 3	0.37	0.36	−0.05	0.39	0.70	−0.3
TM 4	0.32	0.06	0.90	0.17	−0.24	−0.04
TM 5	0.69	−0.48	−0.07	−0.52	0.17	−0.01
TM 7	0.42	−0.20	−0.39	0.63	−0.50	0.02

East Ireland dataset statistics

Variance–covariance matrix

	TM 3	TM 4	TM 5	L band (HH)	L band (VV)	C band (VV)
TM 3	64	−110	86	67	57	76
TM 4		563	−8	−494	−368	−419
TM 5			247	−36	11	−13
L (HH)				3,520	3,191	2,847
L (VV)					3,829	3,303
C (VV)						4,461

Eigenvalues

	PC1	PC2	PC3	PC4	PC5	PC6
	10,243	1,194	591	385	258	11.5
% var.	80.8	9.4	4.7	3.0	2.0	0.1
Cum. % var.	80.8	90.2	94.9	97.9	99.9	100

Eigenvectors

	PC1	PC2	PC3	PC4	PC5	PC6
TM 3	−0.012	0.003	0.17	−0.21	−0.27	−0.92
TM 4	0.08	−0.09	−0.81	0.47	−0.27	−0.18
TM 5	0.002	−0.007	0.02	−0.35	−0.87	0.33
L (HH)	−0.54	0.58	0.25	0.52	−0.21	−0.01
L (VV)	−0.58	0.25	−0.48	−0.57	0.21	−0.01
C (VV)	−0.60	−0.77	0.13	0.15	−0.05	0.008

Nevada dataset statistics

Variance–covariance matrix

	Band 1	Band 3	Band 4	Band 5	Band 7
Band 1	1,310	1,193	384	1,374	1,106
Band 3		1,125	349	1,316	1,044
Band 4			618	721	419
Band 5				2,147	1,524
Band 7					1,154

Eigenvalues

	PC1	PC2	PC3	PC4	PC5
	5,497	563	260.8	21.2	10.8
% var.	86.53	8.86	4.11	0.33	0.17
Cum. % var.	86.53	95.39	99.5	99.83	100

Eigenvectors

	PC1	PC2	PC3	PC4	PC5
TM 1	0.46	−0.42	0.44	−0.481	−0.428
TM 3	0.429	−0.374	0.267	0.716	0.301
TM 4	0.195	0.738	0.621	−0.035	0.172
TM 5	0.604	0.370	−0.49	0.209	−0.463
TM 7	0.451	−0.032	−0.32	−0.459	0.694

Practical Exercise 7
Supervised and unsupervised classification

This practical illustrates two different methods of classification: supervised classification using the Minimum Distance to Mean and unsupervised classification which involves a 'clustering approach'. You should be familiar with the terminology and concepts discussed in section 3.10 before proceeding further. You should also be fully familiar with the DRAGON environment.

The objective of classification is to categorise the pixels of an image into discrete classes or themes automatically. Multispectral data are generally used for classification and the spectral pattern of each pixel within the bands is used as the basis for the categorisation. The DRAGON software included with this book allows the use of four input bands in the classification procedure, and produces up to a maximum of seven classes. The full version of DRAGON allows 16 classes to be produced. This practical should take approximately three hours to complete.

Supervised Classification: Minimum Distance to Mean

Produce a false colour composite with a linear stretch as described in earlier practicals using:

D:\images\May3.img displayed in blue
D:\images\May5.img displayed in green
D:\images\May4.img displayed in red.

The first stage in the supervised classification process is the production of training areas.

- After displaying the image, press ESC twice to go back to the DRAGON Main Menu.
- Click C (Classify images).
- Click TRA (Training signature creation) in the new menu.
- Click DEF (Define training areas on image).
- You are now required to enter the number of the

bands you want to use to define the training areas, three in this case.
- Now input the names of the three image files in the same order as the files used to create the false colour composite image.

   Band 1 image file: D:\images\may3.img
   Band 2 image file: D:\images\may5.img
   Band 3 image file: D:\images\may4.img

Skip band 4 image file.
- Rename the polygon save file: C:\dragon\Mayn.ply after deleting the default.
- Note that the last line of the screen displays =C. This makes the false colour composite stored in the computer's memory the background image.
- Click RUN. The false colour composite is displayed and you are now required to enter the name for signature 1. Type 'Water' and then press RETURN.
- Instructions now appear at the top of the screen. Move the crosshairs (by moving the mouse) over the black water body near the right-hand edge of the screen (south of Leixlip) and press the letter Z on the keyboard. This causes the image to zoom in. Press Z again. DRAGON creates simple circular training areas and you need to input the centre of the circle and also the radius.
- Follow the instruction at the top of the screen. Centre the crosshairs on the water, then press '.' for the centre of the circle. Then move the crosshairs slightly and press '.' again so that an area of water is within the circle. Once the circle is drawn you are asked whether you wish to keep it. If the circle encompasses regions from different surface types then it would not be a suitable training area. If you are happy with it, then press Y and then RETURN to get another training area for the 'water' class. If you do not want to keep this training area, press N and repeat the step.
- Press D to unzoom and then press key F1 to move to another class once you have enough training areas for water.
- You are required to type the number of the class (2), press RETURN and then enter the name of the class (oil seed rape). One very distinctive field

of oil seed rape is shown as a rectangular pink/purple region, south of centre and west of Celbridge. Repeat the above instructions to provide a training region for this surface class.

You should continue this process using Table M1 as a guide to the types of surface which are associated with different colours on the false colour composite. It is important that you take as large a number of training areas as possible for each surface class. Line and pixel co-ordinates are shown for the classes. You can identify these pixels accurately by using the MEASURE option in the GEOMETRY module. (For filename use =C and leave other options at default).

- When you have finished inputting all the classes and defined the training areas, press the F1 key.
- Now you are asked whether you want to continue and build the statistics, answer yes by typing Y and then press RETURN.
- After the calculations have been completed, you are returned to the TRA submenu. Now would be a good time to save the signatures to a file.
- Click R to return to the previous screen and choose EDI to edit the training signatures. The signature file to edit is the current one, so simply click RUN.
- This bring you into an Edit Signatures submenu; pick SAV.

Table M1 Signatures of land classes in study area (East Ireland dataset)

Class	Line/pixel	Signature on false colour composite image
1 Water	207/312	Black
2 Oil seed rape	200/319	Pink/purple
3 Late crops	59/348; 200/307	Greenish/brown
4 Bare soil	53/239	Cyan/light blue
5 Urban/industrial	202/303	Blue
6 Forest/early crops	126/162	Brown/red brown
7 Pasture	199/319	Red/orange

- Save the file by inputting a name including path (e.g. c:\dragon\images\Mayn) and clicking on RUN. The extension .sig will be added automatically.
- Once saved, press ESC to go back to the Classification menu and click MDM (Minimum Distance to Mean method) as the algorithm for the supervised classification method.
- You are required to enter some parameters, enter 3 as the number of bands you intend to use and press RETURN.
- Enter again the names of the image files by typing or selecting them. The image files are: D:\images\May3.img, D:\images\May5.img and D:\images\May4.img. Follow this order. Press RETURN at the end.
- We will use the current signature file, so leave this unchanged.
- You are now required to enter a threshold number between 1 and 10. The larger the number picked, the fewer pixels will be unclassified. For now choose a value of 2.
- Do not print the class table.
- Click on RUN at the bottom right of the screen to start the classification and display the results.
- The screen will display the Classification Statistics, which you should note down. Then click on OK to display the classified image.
- Press ESC to return to the Classification menu. The classified image stored in memory can now be saved.
- Press ESC to go back to the DRAGON Main Menu.
- Click F (File Handling operations) to save the results of the classification
- Click SAV to save the image currently stored in the computer's memory.
- Give the file a name (such as c:\dragon\images\mayn.img) and press ENTER.
- If desired you can add a descriptive comment to the image file. Type it in the appropriate line and then click on RUN.
- You are also asked whether you want the class names to be added, type N then press RETURN. In the full version of DRAGON, you can create a

Table M2 Classification statistics for East Ireland dataset

Class	No. of pixels	% area	Colour (after classification)
0 No class	30,321	18.95	black
1 Water	171	0.10	blue
2 Oil seed rape	97	0.06	green
3 Late crops	73,077	45.67	light blue
4 Bare soil	6,206	3.87	red
5 Urban/industrial	14,762	9.22	purple
6 Forest/early crops	12,767	7.97	orange
7 Pasture	22,599	14.12	grey

legend which can be displayed beside the image. This procedure is not available on the version of DRAGON included here. This image can be viewed using the 1BA image colour display option from the Display menu.

- Repeat the above Minimum Distance to Mean classification procedure but instead of using the current signature file, use D:\images\maysup.sig. Again, pick a value of 2 for the threshold. You should obtain Table M2.

Click on the OK at the bottom of the screen when the statistics have been examined.

The computer will display the classified image. (If you were unable to produce this image, it is included within the East Ireland dataset. It is termed Maysup. img. View it by using the 1BA image-colour display option from the Display menu.)

Describe the image.

In general the classification has been reasonably accurate. Soils, shown red have been picked out and the small amount of oil seed rape (green) and water (dark blue) in the scene has also been identified. There has been some misclassification of forested areas. A purple fringe around the northern and eastern margins of Carton estate is a forested region but it is shown as urban on the classified image. A large percentage of the unclassified areas represents field boundaries. The percentage unclassified could be reduced if training areas were acquired which encompassed the field boundaries.

Editing the Training Areas

After viewing the classified image, you may wish to edit your own training areas. For example, you may decide that you want to increase the number of training areas for a specific class. To do this:

- First redisplay the original false colour composite image using the 3BA routine.
- Then go into the Classify Images menu and click on Training signature creation.
- Now click on Apply stored training area boundaries.
- You are now required to enter the number of the bands you want to use to define the training areas, three in this case.
- Now input the names of the three image files in the same order as the files used to create the false colour composite image.

      Band 1 image file: D:\images\may3.img
      Band 2 image file: D:\images\may5.img
      Band 3 image file: D:\images\may4.img

- The polygon file to apply is the one you initially used: C: \dragon\images\Mayn.ply.
- Click RUN and, as before, build the statistics by typing Y then press RETURN.
- Choose DEF in Training Signature Creation and complete the parameters as before, but for the background image, type =C, which was the last colour image displayed and click on RUN. (You will be warned that the polgon file C: \dragon\images\Mayn.ply already exists.)

- The original image is displayed and you are asked whether you wish to keep the old circles. If you are happy with them type Y and then press RETURN.
- You are now required to pick a class to edit. Whichever class you pick, the original training areas are displayed. You may now add others using the procedures discussed earlier. When you have completed your changes you may finish by pressing the key F1.
- From the EDIT signatures facility you also have the option to delete, modify and combine signatures.

Unsupervised Classification

The objective of unsupervised classification is to group digital numbers obtained in several bands into clusters based on the statistical characteristics of the digital numbers. Thus, a small range of digital numbers (DNs), in our sample 3 band image may constitute one cluster that is distinct from another range combination which represents another cluster. As explained in section 3.10, it is important to realise that the clusters represent spectral classes and not information classes.

First, display the following bands with a simple linear contrast stretch:

D:\images\May3.img displayed in blue
D:\images\May5.img displayed in green
D:\images\May4.img displayed in red.

Then:

- Press ESC twice to go back to the DRAGON Main Menu.
- Click C (Classify images).
- Click CLU (Clustering method).
- Enter 3 as the number of bands you intend to use, press RETURN.
- Enter again the name of the image files by typing them or selecting them after clicking the blue triangle at the end of the line. Input the files in the order:

D:\images\May3.img
D:\images\May5.img
D:\images\May4.img.

- By default the DRAGON software displays the maximum number possible of clusters (7); leave unchanged. Also leave the number of repetitions unchanged (10). (If you want the program to execute all the repetitions without prompting then type 11 for the number of repetitions between questions.)
- Click RUN at the bottom left of the screen.
- The screen will display the clustering results after every repetition, asking whether you want to stop clustering. Answer NO by typing N then press RETURN.
- Note that, at every iteration, the number of pixels grouped in the clusters changes.
- At the end of the ten repetitions the screen displays the 'clustering statistics'. Make a note of these before clicking on OK. You should get the results shown in Table M3.

Click on OK after making a note of the figures. The screen now displays the unsupervised classified image. You can save the image resulting from clustering as described earlier but do not enter class names. The unsupervised procedure classifies all the pixels.

Briefly compare the unsupervised image with the supervised one created earlier.

The image shows the oil seed rape field as a distinctive bright red colour (class 4) though the area is greater than

Table M3 Statistics for unsupervised classification of East Ireland dataset

Class/cluster	No. of pixels	% area	Composition
Class 1	30,145	18.84	1
Class 2	49,243	30.77	2
Class 3	12,096	7.56	3
Class 4	194	0.12	4
Class 5	34,571	21.60	5
Class 6	19,295	12.05	6
Class 7	14,456	9.03	7

that calculated using the supervised classification. While water was a separate class in the supervised classification, it is shown grey on the unsupervised classified image (class 7), similar to urban areas. Some areas which are known to be forested are also displayed in grey. The pale blue has picked out many of the fields with bare soils (class 3) though a comparison with the soils spectral class in the supervised image shows that some of the unclassified pixels have been placed in this class. Pasture (class 5) is shown in purple and again some of the unclassified pixels have been put in this class. Orange regions (class 6) accord mainly with early crops and forests. Class 1 (dark blue) and class 2 (green) coincide with the late crops' fields in the supervised image. This illustrates how supervised and unsupervised classification can yield different results. The training areas for the late crops encompassed crops at different stages of their phenolog-ical cycle (thus they are associated with different spectral signatures) whereas the unsupervised classification placed these fields, with a range of spectral signatures, into two separate classes.

You may wish to attempt to classify the East Ireland dataset using both Landsat TM and radar images as the inputs. Various land-cover classes have been discussed in earlier practicals for some of the image datasets. You should use this information to perform a supervised classification of these images. Unsupervised classification can be performed on all the datasets. Remember that the bands you employ should not be highly correlated. You may use the scatterplot routine to determine the extent of correlation or even choose to classify the principal component images.

INDEX

Note: Figures and Tables are indicated (in this index) by *italic page numbers*, and Boxes by **bold page numbers**; "*Pl.*" means "Colour Plate ..." and "RS" = "remote sensing"

CD-ROM USER LICENCE AGREEMENT

We welcome you as a user of this Routledge CD-ROM and hope that you find it a useful and valuable tool. Please read this document carefully. **This is a legal agreement** between you (hereinafter referred to as the "Licensee") and Routledge (the "Publisher") which defines the terms under which you may use the Product. By opening the package containing the CD-ROM you have agreed to these terms and conditions outlined herein. If you do not agree to these terms you must return the Product to Routledge intact with all its components as listed on the back of the package, within ten days of purchase and the purchase price will be refunded to you.

1. Definition of the Product

The product which is the subject of this Agreement, the *Introductory Remote Sensing: Digital Image Processing and Applications* CD-ROM (the "Product") consists of:

1.1   Underlying data comprised in the product (the "Data")

1.2   A compilation of the Data (the "Database")

1.3   Software (the "Software") for accessing and using the Database

1.4   User guide (the "Manual")

1.5   A CD-ROM disk (the "CD-ROM")

2. Commencement and Licence

2.1   This Agreement commences upon the breaking open of the package containing the CD-ROM by the Licensee (the "Commencement Date").

2.2   This is a licence agreement (the "Agreement") for the use of the Product by the Licensee, and not an agreement for sale.

2.3   The Publisher licenses the Licensee on a non-exclusive and non-transferable basis to use the Product on condition that the Licensee complies with this Agreement. The Licensee acknowledges that it is only permitted to use the Product in accordance with this Agreement.

3. Installation and Use

3.1   The Licensee may provide access to the Product only on a single personal computer for individual study. Multi-user use or networking is only permissible with the express permission of the Publisher in writing and requires payment of the appropriate fee as specified by the Publisher, and signature by the Licensee of a separate multi-user licence agreement.

3.2   The Licensee shall be responsible for installing the Product and for the effectiveness of such installation.

4. Permitted Activities

4.1   The Licensee shall be entitled:

4.1.1   to use the Product for its own internal purposes;

4.1.2   to download onto electronic, magnetic, optical or similar storage medium reasonable portions of the Database provided that the purpose of the Licensee is to undertake internal research or study and provided that such storage is temporary;

4.1.3   to make a copy of the Database and/or the Software for back-up/archival/disaster recovery purposes only.

4.2   The Licensee acknowledges that its rights to use the Product are strictly as set out in this Agreement, and all other uses (whether expressly mentioned in Clause 5 below or not) are prohibited.

5. Prohibited Activities

The following are prohibited without the express permission of the Publisher:

5.1   The commercial exploitation of any part of the Product.

5.2   The rental, loan (free or for money or money's worth) or hire purchase of the product, save with the express consent of the Publisher.

5.3   Any activity which raises the reasonable prospect of impeding the Publisher's ability or opportunities to market the Product.

5.4   Any provision of services to third parties using the Product, whether by way of trade or otherwise.

5.5   Any networking, physical or electronic distribution or dissemination of the product save as expressly permitted by this Agreement.

5.6   Any reverse engineering, decompilation, dis-

assembly or other alteration of the Software save in accordance with applicable national laws.

5.7   The right to create any derivative product or service from the Product save as expressly provided for in this Agreement.

5.8   The use of the Software separately from the Database.

5.9   Any alteration, amendment, modification or deletion from the Product, whether for the purposes of error correction or otherwise.

5.10   The merging of the Database or the Software with any other database or software.

5.11   Any testing, study or analysis of the Software save to study its underlying ideas and principles.

6. General Responsibilities of the Licensee

6.1   The Licensee will take all reasonable steps to ensure that the Product is used in accordance with the terms and conditions of this Agreement.

6.2   The Licensee acknowledges that damages may not be a sufficient remedy for the Publisher in the event of breach of this Agreement by the Licensee, and that an injunction may be appropriate.

6.3   The Licensee undertakes to keep the Product safe and to use its best endeavours to ensure that the product does not fall into the hands of third parties, whether as a result of theft or otherwise.

6.4   Where information of a confidential nature relating to the product or the business affairs of the Publisher comes into the possession of the Licensee pursuant to this Agreement (or otherwise), the Licensee agrees to use such information solely for the purposes of this Agreement, and under no circumstances to disclose any element of the information to any third party save strictly as permitted under this Agreement. For the avoidance of doubt, the Licensee's obligations under this sub-clause 6.4 shall survive termination of this Agreement.

7. Warrant and Liability

7.1   The Publisher warrants that it has the authority to enter into this Agreement, and that it has secured all rights and permissions necessary to enable the Licensee to use the Product in accordance with this Agreement.

7.2   The Publisher warrants that the CD-ROM as supplied on the Commencement Date shall be free of defects in materials and workmanship, and undertakes to replace any defective CD-ROM within 28 days of notice of such defect being received provided such notice is received within 30 days of such supply. As an alternative to replacement, the Publisher agrees fully to refund the Licensee in such circumstances, if the Licensee so requests, provided that the Licensee returns the Product to the Publisher. The provisions of this sub-clause 7.2 do not apply where the defect results from an accident or from misuse of the product by the Licensee.

7.3   Sub-clause 7.2 sets out the sole and exclusive remedy of the Licensee in relation to defects in the CD-ROM.

7.4   The Publisher and the Licensee acknowledge that the Publisher supplies the Product on as "as is" basis. The Publisher gives no warranties:

7.4.1   that the Product satisfies the individual requirements of the Licensee; or

7.4.2   that the product is otherwise fit for the Licensee's purpose; or

7.4.3   that the Data is accurate or complete or free of errors or omissions; or

7.4.4   that the product is compatible with the Licensee's hardware equipment and software operating environment.

7.5   The Publisher hereby disclaims all warranties and conditions, express or implied, which are not stated above.

7.6   Nothing in this Clause 7 limits the Publisher's liability to the Licensee in the event of death or personal injury resulting from the Publisher's negligence.

7.7   The Publisher hereby excludes liability for loss of revenue, reputation, business, profits, or for indirect or consequential losses, irrespective of whether the Publisher was advised by the Licensee of the potential of such losses.

7.8   The Licensee acknowledges the merit of independently verifying Data prior to taking any decisions of material significance (commercial or otherwise) based on such data. It is agreed that the Publisher shall not be liable for any losses which result from the Licensee placing reliance on the Data or on the database, under any circumstances.

7.9   Subject to sub-clause 7.6 above, the Publisher's liability under this Agreement shall be limited to the purchase price.

8. Intellectual Property Rights

8.1   Nothing in this Agreement affects the ownership of copyright or other intellectual property rights in the Data, the Database, the Software or the Manual.

8.2   The Licensee agrees to display the Publisher's copyright notice in the manner described in the Manual and in the Product.

8.3   The Licensee hereby agrees to abide by copyright and similar notice requirements required by the Publisher, details of which are as follows: "(c) 2000 Routledge and their licensors. All Rights Reserved. All materials in the *Introductory Remote Sensing: Digital Image Processing and Applications* CD-ROM and accompanying User Guide are copyright protected. No such materials may be used, displayed, modified, adapted, distributed, transmitted, transferred, published or otherwise reproduced in any form or by any means now or hereafter developed other than strictly in accordance with the terms of the licence agreement enclosed with the CD-ROM. However, text may be printed and copied for research and private study within the preset program limitations. Please note the copyright notice above, and that any text or images printed or copied must credit the source."

8.4   This Product contains material proprietary to and copyrighted by the Publisher and others. Except for the licence granted herein, all rights, title and interest in the Product, in all languages, formats and media throughout the world, including all copyrights therein, are and remain the property of the Publisher or other copyright owners identified in the Product.

9. Non-assignment

This Agreement and the licence contained within it may not be assigned to any other person or entity without the written consent of the Publisher.

10. Termination and Consequences of Termination

10.1   The Publisher shall have the right to terminate this Agreement if:

10.1.1   the Licensee is in material breach of this Agreement and fails to remedy such breach (where capable of remedy) within 14 days of a written notice from the Publisher requiring it to do so; or

10.1.2   the Licensee becomes insolvent, becomes subject to receivership, liquidation or similar external administration; or

10.1.3   the Licensee ceases to operate in business.

10.2   The Licensee shall have the right to terminate this Agreement for any reason upon two months' written notice. The Licensee shall not be entitled to

any refund for payments made under this Agreement prior to termination under this sub-clause 10.2.

10.3   Termination by either of the parties is without prejudice to any other rights or remedies under the general law to which they may be entitled, or which survive such termination (including rights of the Publisher under sub-clause 6.4 above).

10.4   Upon termination of this Agreement, or expiry of its terms, the Licensee must:

10.4.1   destroy all back up copies of the product; and

10.4.2   return the Product to the Publisher.

11. General

11.1   *Compliance with export provisions*
The Publisher hereby agrees to comply fully with all relevant export laws and regulations of the United Kingdom to ensure that the Product is not exported, directly or indirectly, in violation of English law.

11.2   *Force majeure*
The parties accept no responsibility for breaches of this Agreement occurring as a result of circumstances beyond their control.

11.3   *No waiver*
Any failure or delay by either party to exercise or enforce any right conferred by this Agreement shall not be deemed to be a waiver of such right.

11.4   *Entire agreement*
This Agreement represents the entire agreement between the Publisher and the Licensee concerning the Product. The terms of this Agreement supersede all prior purchase orders, written terms and conditions, written or verbal representations, advertising or statements relating in any way to the Product.

11.5   *Severability*
If any provision of this Agreement is found to be invalid or unenforceable by a court of law of compe-

tent jurisdiction, such a finding shall not affect the other provisions of this Agreement and all provisions of this Agreement unaffected by such a finding shall remain in full force and effect.

11.6 *Variations*
This Agreement may only be varied in writing by means of variation signed in writing by both parties.

11.7 *Notices*
All notices to be delivered to: Routledge, 11 New Fetter Lane, London EC4P 4EE, UK.

11.8 *Governing law*
This Agreement is governed by English law and the parties hereby agree that any dispute arising under this Agreement shall be subject to the jurisdiction of the English courts.

If you have any queries about the terms of this licence, please contact:

Routledge
11 New Fetter Lane
London EC4P 4EE
Tel: (0)20 7583 9855
Fax: (0)20 7842298
Email: info@tandf.co.uk

SPECIAL OFFER FOR READERS OF
INTRODUCTORY REMOTE SENSING°

You can get a copy of the full DRAGON/ips® Academic Edition remote sensing software system for a fraction of the normal price.

DRAGON Academic Edition provides image processing capabilities far beyond what you have seen in the restricted version of DRAGON included with this book.

Features include:

- Larger image capacity (1,024 × 1,024)
- Full color, multi-image display and annotation
- Image import, export, capture and printing
- Supervised and unsupervised classification (clustering, maximum likelihood, etc.)
- Classification accuracy assessment
- Geometric correction and image-to-image registration
- On screen vector capture
- Principal components analysis
- GIS integration functions
- Both Microsoft Windows™ and MS DOS versions
      and much, much more!   (See the CD-ROM for detailed technical information.)

All this for only: US$ 350 (Windows and MS-DOS)
US$ 250 (MS-DOS only)

This represents more than a 60% discount off usual prices!

To take advantage of this offer, copy the coupon on the reverse, or print the order form from the CD-ROM, and send with payment to:

Goldlin-Rudahl Systems, Inc.
PMB 213, 6 University Drive, Suite 206   FAX: +1-413-549-6401
Amherst, MA 01002   email: info@goldin-rudahl.com
U.S.A.   WWW: http://www.goldin-rudahl.com

In the U.K., you may contact:

I.S. Ltd.
Atlas House, Atlas Business Centre   Phone: 0161-499-7609
Simonsway   FAX: 0161-436-6690
Manchester M22 5HF   email: isman@compuserve.com

° Limited time offer: Discount prices guaranteed until January 1, 2002

Instructors: Please contact Goldin-Rudahl Systems for information on discounts on quantity purchases, the DRAGON Professional Suite, site licences, and other products.

Order Form

Please send me:

- [ ] DRAGON/ips® Academic Edition for Windows and MS-DOS (US$ 350)*
- [ ] DRAGON/ips® Academic Edition for MS-DOS (US$ 250)
(Prices include shipping by air post.)

Payment type:

- [ ] Cheque payable by a U.S. Bank
- [ ] Visa
- [ ] Mastercard

Credit card number: _____ Expires: _____

Authorized signature: _____

Ship to:

Name: _____

Address: _____

_____

City: _____ State/Province/County: _____

Postal code: _____ Country: _____

Email: _____

Fax: _____

Mail to:

Goldin-Rudahl Systems, Inc.
PMB 213, 6 University Drive, Suite 206, Amherst, MA 01002 USA

Fax to:

(Credit card holders only)
Goldin-Rudahl Systems, Inc.
+1-413-549-6401

* Windows version requires at least 32,000-color gaphics (15 bit color) 32 MBytes memory and CD-ROM drive.
  MS-DOS version requires at least VGA graphics and 2 MBytes memory.